COUPLED NONLINEAR OSCILLATORS

NORTH-HOLLAND
MATHEMATICS STUDIES **80**

COUPLED NONLINEAR OSCILLATORS

Proceedings of the Joint U.S. Army-
Center for Nonlinear Studies Workshop,
held in Los Alamos, New Mexico, 21-23 July, 1981

Edited by
J. CHANDRA
U.S. Army Research Office
Research Triangle Park
North Carolina, U.S.A.

and

A. C. SCOTT
Center for Nonlinear Studies
Los Alamos National Laboratory
Los Alamos, New Mexico, U.S.A.

1983

NORTH-HOLLAND PUBLISHING COMPANY · AMSTERDAM · NEW YORK · OXFORD

ISBN: 0 444 86677 9

Publishers:
NORTH-HOLLAND PUBLISHING COMPANY
AMSTERDAM · NEW YORK · OXFORD

Sole distributors for the U.S.A. and Canada:
ELSEVIER SCIENCE PUBLISHING COMPANY, INC.
52 VANDERBILT AVENUE, NEW YORK, N.Y. 10017

Library of Congress Cataloging in Publication Data
Main entry under title:

Coupled nonlinear oscillators.

(North-Holland mathematics studies ; 80)
1. Nonlinear oscillators--Congresses. 2. Chaotic behavior in systems--Congresses. 3. Josephson junctions --Congresses. 4. Lasers--Congresses. 5. Solitons-- Congresses. 6. Bifurcation theory--Congresses.
I. Chandra, J. II. Scott, Alwyn, 1931- .
III. United States. Army. IV. Center for Nonlinear Studies (Los Alamos National Laboratory) V. Series.
TK7872.07C65 1983 621.3815'33 83-11427
ISBN 0-444-86677-9 (U.S.)

PRINTED IN THE NETHERLANDS

PREFACE

These are exciting times for applied mathematicians. No longer are their results placed on the back shelves of libraries to await the attention of readers from following generations; instead their newly invited ideas are being used in experimental science. A particularly striking example of this phenomena is in the general area of "chaos". Chaos can be described as sensitive dependence on initial conditions, associated with erratic motion on the energy hypersurfaces in conservative systems. Except in a general sense, the word was unknown ten years ago. Now one finds scientists in such diverse areas of research as nonlinear optics, laser technology, Josephson (superconductive) oscillators, hydrodynamic turbulence, and chemical and biological oscillators are approaching their work with new perspectives.

Recognizing the importance of these developments the Mathematics Division, U. S. Army Research Office, and the Center for Nonlinear Studies, Los Alamos National Laboratory, jointly sponsored a workshop on "Coupled Nonlinear Oscillators" which was held at the Los Alamos National Laboratory on 21-23 July 1981. The workshop was interdisciplinary drawing participants from the ranks of applied mathematicians and applied scientists from such fields as laser optics, hydrodynamics, and electrical engineering. The workshop was exciting and productive. In the words of one participant it blended mathematical techniques from bifurcation theory and chaotic motion of dynamical systems with the theoretical and experimental results of dispersive optical bistability, coherence effects, absorption dynamics of multi-level systems and soliton motion in Josephson junction oscillators.

The aim of this book is to record the spirit of the workshop without being a precise compendium of the papers presented. Thus, we are including a selected subset of those papers plus some additional invited contributions which we feel will allow the reader to gain an understanding and appreciation of this important research area.

A knowledge of the behavior of idealized nonlinearly-coupled oscillators can provide fundamental understanding of the behavior of more complicated discrete or continuous systems. These oscillators display transitions to quasi periodic states, reverse bifurcation to phase-locked periodic states and chaotic behavior. In Chapter I, Steen and Davis consider transitions in weakly-coupled nonlinear oscillators. Chow and Mallet-Paret, Chapter II, consider a class of singularly perturbed delay-differential equation which describes an optically bistable device. They investigate chaotic behavior for certain parametric values. Earlier these were predicated numerically and confirmed experimentally by others. Haken-Lamb theory of lasers reduces the dynamics to be studied to that of one or several coupled nonlinear oscillators, classical or quantized. Indeed, phenomena familiar from nonlinear oscillators like limit-cycles, frequency pulling, phase and frequency locking between various modes, combination tones are all commonly observed in laser systems. However, an additional feature of nonlinear oscillators has not yet received much attention in study of lasers, namely, the ability of nonlinear oscillators for chaotic motion. In Chapter III, Graham considers various cases where chaos appears in simple laser systems. The truncated Navier-Stokes equations have proven to be interesting mathematical models of nonlinear differential equations exhibiting chaotic behavior. Such models show the important role of period-doubling bifurcations of periodic orbits as pathways to turbulence in multi-dimensional dissipative systems. In Chapter IV, Franceschini investigates general solutions of a truncated Navier-Stokes equation on a two dimensional torus.

The next two chapters consider oscillating phenomena in Josephson Junctions. In Chapter V, Hagan employes singular perturbation techniques to obtain expressions for the response of a point junction which are accurate even when resonance occurs. Lomdahl, Soerensen and Christiansen in Chapter IV undertake a detailed numerical study of a Sine-Gordon model of the Josephson tunnel junction and compare with experimental observations with different length-to-Josephson-penetration depth ratios. The soliton excitations in Josephson tunnel junctions studied in this chapter continue to draw considerable attention.

The coupled Maxwell-Schrödinger equations predict a variety of fascinating and well-studied propagation phenomena including self-induced transparency, pulse reshaping, and pulse breakup into solitons. These effects are associated with the simplest (two-level) model of molecular absorber. In Chapter VII, Shore and Eberly, report several results on three models proposed as descriptive of pulse propagation through molecular media. The most important features of the Jaynes-Cummings model are that it is fully quantum mechanical and exactly soluble. The solutions are, therefore, not restrictive to domains of small coupling constant or weak excitation and are not subject to approximate decorrelations or truncations of a mean-field type. In Chapter VIII, Hioe, Yoo and Eberly are concerned with the properties of dynamics of the Jaynes-Cummings model in certain "irregular" regime. In the final chapter, Bowden and Eberly consider some dynamic aspects of interrupted coarse-graining in stimulated excitation of vibronic bands.

We thank all of those whose efforts have helped both to make the workshop successful and to bring this book into its final form. We particularly appreciate the help of Professor Donald Drew who, as one of the main organizers of the workshop was a moving force from the beginning. We are indebted to Mrs. Marian Martinez for excellent typing of the proceedings.

Special thanks are due to Mrs. Sherry Duke of the Army Research Office and Mrs. Janet Gerwin of Los Alamos National Laboratory for their diligent cooperation through all phases of the workshop.

November 1982

Jagdish Chandra

Alwyn C. Scott

CONTENTS

COUPLED NONLINEAR OSCILLATORS
J. Chandra and A.C. Scott (eds.)

TRANSITIONS IN WEAKLY-COUPLED NONLINEAR OSCILLATORS

Paul H. Steen[(a)] and Stephen H. Davis

Department of Engineering Sciences and Applied Mathematics
Northwestern University
Evanston, Illinois 60201

Two oscillators having natural frequencies δ and γ and damping coefficients proportional to $-\lambda$ are weakly coupled through cubic nonlinearities. As λ is increased through $\lambda_c \equiv 0$, nonlinear solutions of amplitude ε bifurcate. To leading order in ε such systems (including the case of coupled van der Pol oscillators) have behaviors governed by the single parameter $\rho \sim 2(\delta-\gamma)/(\lambda-\lambda_c)$. ρ has a critical value ρ_c determined by a Josephson-junction equation. For $|\rho| > \rho_c$ the oscillations are quasiperiodic; the modulations die out in the far-from-resonance case $|\rho| \to \infty$. For $|\rho| < \rho_c$ there are phase-locked oscillations and competing stable states can occur. An increase in λ, causing $|\rho|$ to cross ρ_c, results in a type of reverse bifurcation (quasiperiodic to periodic). No chaotic behavior is found.

Complex physical systems often display orderly sequences of transitions among states. Simplified models of rudimentary systems can often shed light on the physics of such transitions. Hence, there is great interest [1] in the predictions of highly idealized difference equations. Experimental observations in fluid dynamics [2], chemical [3] and laser [4] systems commonly reveal transitions to quasiperiodic states. The lack of direct theoretical results on such behavior leads us to pose a model system consisting of a pair of nonlinearly coupled oscillators.

$$\frac{d^2x}{dt^2} - \lambda \frac{dx}{dt} + \delta^2 x = f\left(x, \frac{dx}{dt}, y, \frac{dy}{dt}\right), \tag{1a}$$

$$\frac{d^2y}{dt^2} - \lambda \frac{dy}{dt} + \gamma^2 y = g\left(x, \frac{dx}{dt}, y, \frac{dy}{dt}\right) \tag{1b}$$

where δ and γ are the linear-oscillator frequencies at zero damping, and the nonlinear functions f and g vanish for $x = y = 0$. The damping coefficient $-\lambda$ is the bifurcation parameter in that the linearized oscillators have decaying solutions for $\lambda < 0$ and growing solution for $\lambda > 0$. At the critical value $\lambda_c = 0$ non-trivial solutions may bifurcate from the null solution.

We can generalize system (1) slightly by considering the following system of four first-order equations:

$$\frac{d}{dt}\begin{bmatrix} x \\ u \\ y \\ v \end{bmatrix} - \lambda \begin{bmatrix} b_1 & 0 & 0 & 0 \\ 0 & b_2 & 0 & 0 \\ 0 & 0 & b_3 & 0 \\ 0 & 0 & 0 & b_4 \end{bmatrix} \begin{bmatrix} x \\ u \\ y \\ v \end{bmatrix} + \begin{bmatrix} 0 & -\delta & 0 & 0 \\ \delta & 0 & 0 & 0 \\ 0 & 0 & 0 & -\gamma \\ 0 & 0 & \gamma & 0 \end{bmatrix} \begin{bmatrix} x \\ u \\ y \\ v \end{bmatrix} = \underset{\sim}{C} \tag{2}$$

where the four-vector $\underset{\sim}{C}$ depends on x, u, y and v. All of the above discussed properties remain. A special case of system (2) is a pair of coupled van der Pol oscillators. In order to bypass certain mathematical technicalities we confine attention to <u>nonlinearities that are cubic and satisfy a strong-definiteness condition</u>,[b] viz. there exists a positive constant K such that

$$(x,u,y,v)\cdot\underset{\sim}{C} < -K(x^2+u^2+y^2+v^2)^2 \; . \tag{3}$$

Condition (3) ensures that all solutions of system (2) will remain in a bounded neighborhood of the origin. We also assume that both δ and γ are bounded away from zero so that the only strong resonance for weakly nonlinear interactions occurs near $\delta = \gamma$.

We seek asymptotic solutions in terms of an amplitude parameter ε as follows:

$$x(t;\varepsilon) \sim \varepsilon i A(T_2) e^{i\delta T_0} + O(\varepsilon^2) + \text{complex conjugate} \tag{4a}$$

$$y(t;\varepsilon) \sim \varepsilon i B(T_2) e^{i\delta T_0} + O(\varepsilon^2) + \text{complex conjugate} \tag{4b}$$

$$\lambda(\varepsilon) \sim \lambda_c + \varepsilon^2\lambda_2 + O(\varepsilon^3) \tag{4c}$$

by using a two time-scale expansion where

$$T_0 \equiv t \quad \text{and} \quad T_2 \equiv \varepsilon^2 t \; . \tag{4d}$$

In order to describe the strong resonance near $\delta = \gamma$, we relate [5] the two frequencies as follows:

$$\delta - \gamma \equiv \frac{1}{2}k\varepsilon^2 \; , \; k = O(1) \quad \text{as} \quad \varepsilon \to 0 \; . \tag{5}$$

The forms (4) and (5) are substituted into system (2) and multiple-scale analysis is used. At order ε^3 a solvability condition determines equations for the envelopes $A(T_2)$ and $B(T_2)$. If we use the polar forms $A(T_2) = a(T_2)\exp[i\phi(T_2)]$ and $B(T_2) = b(T_2)\exp[i\psi(T_2)$, the resulting envelope equations are as follows:

$$\frac{d}{dT_2}a^2 = a^2\{\lambda_2 c_1 + c_2 a^2 + (c_3 + c_4 \cos Z)b^2\} \tag{6a}$$

$$\frac{d}{dT_2}b^2 = b^2\{\lambda_2 d_1 + d_2 b^2 + (d_3 + d_4 \cos Z)a^2\} \tag{6b}$$

$$\frac{d}{dT_2}Z = -(d_4 a^2 + c_4 b^2)\sin Z - k \tag{6c}$$

$$\frac{d}{dT_2}\phi = \frac{1}{2} b^2 c_4 \sin Z . \tag{7}$$

Here we have written the phase-difference as follows:

$$Z \equiv 2(\psi - \phi) - kT_2 . \tag{8}$$

The c_i, d_i, $i = 1,\ldots 4$ are computable numbers dependent on the nonlinearity $\underset{\sim}{C}$ that are restricted by condition (3). Condition (3) also implies that $\lambda_2 > 0$. Hence, if $\lambda < 0$ all solutions decay to zero and nontrivial solutions exist only for $\lambda > 0$.

The dominant behavior of the oscillator amplitudes a and b is coupled to the phase difference through Z. Thus, the governing system is three dimensional. The individual phases ϕ and ψ can then be obtained by using Eqs. (7) and (8). It is easy to see that simple rescaling combines the two parameters k and λ_2 into a single parameter ρ

$$\rho \equiv k/\lambda_2 \sim 2(\delta-\gamma)/(\lambda-\lambda_c) . \tag{9}$$

ρ is the entrainment parameter, which is to leading order in ε the ratio of the frequency difference $\delta - \gamma$ to the degree of supercriticality. The behavior of the oscillators (2) is described by system (6) for all frequencies δ and γ since we find that the formal limit $|\rho| \to \infty$ gives the correct behaviors for δ and γ widely separated i.e. for $k \to \infty$ in Eq. (5).

There are three types of solutions of system (6) that are found. There are pure mode ($a \equiv 0$ or $b \equiv 0$) constants, mixed-mode ($a,b \not\equiv 0$) constants and mixed-mode ($a,b \not\equiv 0$) limit cycles. These correspond through Eqs. (4) to the oscillators having pure-mode periodic solutions, mixed-mode periodic solutions and mixed-mode quasiperiodic behavior, respectively. No chaotic solutions of system (6) are found.

There is a critical value ρ_c of $|\rho|$. When $|\rho| < \rho_c$ there are typically four or eight mixed-mode constant solutions at most half of which are stable. These are phase-locked periodic solutions for the oscillators. There can be competing stable attractors. There coexist pure-mode constant solutions whose stability depends on details of $\underset{\sim}{C}$. When $|\rho| > \rho_c$, there are no mixed-mode constant solutions but mixed-mode limit cycles exist and are stable. There coexist pure-mode constant solutions whose stability depends on details of $\underset{\sim}{C}$. The results are depicted in Fig. 1.

Note that the present coupled-oscillator system displays a type of reverse bifurcation prevalent in many physical systems [4]. Here, if $\delta - \gamma$ is fixed and λ is increased, there is a transition at ρ_c from quasiperiodic oscillations to phase-locked periodic oscillations.

There is an abrupt change in behavior as $|\rho|$ crosses ρ_c. The change is controlled by the phase-difference equation (6c) whose qualitative nature is captured by the scalar equation

$$\frac{dZ}{dt} = \sin Z - \rho . \tag{10}$$

Here $\rho_c = 1$. For $|\rho| > \rho_c$ there is a large amplitude periodic solution replacing the pair of steady states that typically exist for $|\rho| < \rho_c$. Equation (10) governs the phase-difference of the wave functions across a Josephson junction [6] when the conductance is large; ρ plays the role of the current.

The detailed asymptotic analysis, the proof of existence of periodic amplitudes for $|\rho| > \rho_c$ and the discussion of the numerical experiments will be covered in a forthcoming paper.

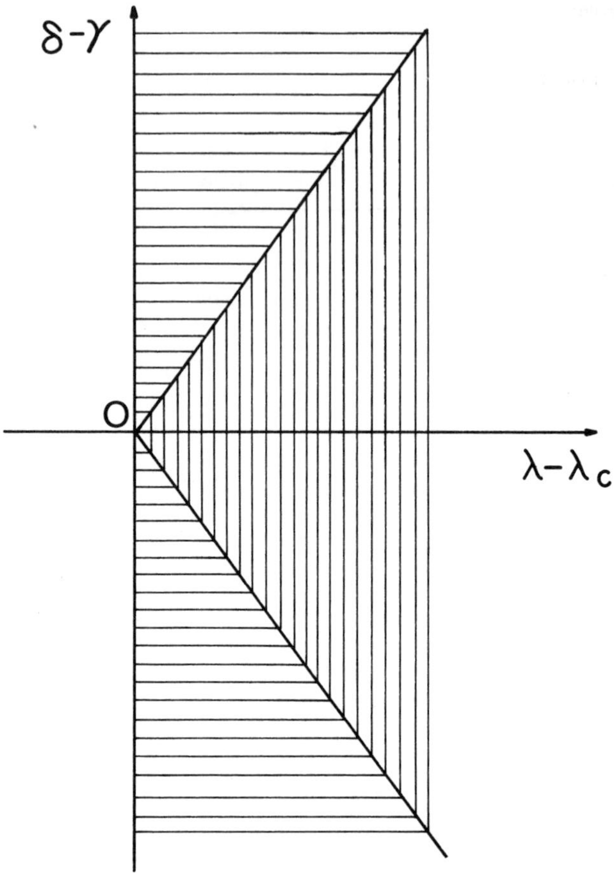

Figure 1
The (δ-γ) vs. (λ-λ_c) plane for small ε. The slopes of the lines are ± 1/2 ρ_c. Outside the rays in the horizontally-hatched region only periodic mixed-mode amplitudes exist. Inside the rays in the vertically-hatched region steady mixed-mode amplitudes exist. To the left in the unhatched region all nontrivial solutions decay to zero.

A knowledge of the behaviors of nonlinearly-coupled oscillators gives one an elemental understanding of the behaviors of more complicated discrete systems as well as some continuous systems. These oscillators display transitions to quasiperiodic states, reverse bifurcations to phase-locked periodic states and the possibility of competing stable phase-locked states.

We wish to thank Professor S. Rosenblat for his constructive advice. This work was supported by grants from the Army Research Office, Applied Mathematics Program.

(a) Current address: Department of Chemical Engineering
Stanford University, Stanford, CA 94305

(b) $\underset{\sim}{C}$ can be defined from a trilinear form each of whose three arguments is (x,u,y,v). The trilinear form is constructed by replacing each of the 40 nonzero entries in the most general fourth-order isotropic tensor by a negative constant.

REFERENCES

1. Physics Today, News-Search and Discovery, <u>34</u>, 17 (March 1981).
2. H. L. Swinney and J. P. Gollub, Physics Today <u>31</u>, 41 (August 1978).
3. M. Marek and I. Stuchl, Biophys. Chem. <u>3</u>, 241 (1975).
4. Bowden, C. M., Ciftan, M. and Robl, H. R., <u>Optical Bistability</u>, Plenum Publ., New York (1981).
5. H. Kabakow, Ph.D. Dissertation, Cal. Tech., 1968; Int. J. Nonl. Mech. <u>7</u>, 125 (1972).
6. C. Kittel, <u>Introduction to Solid State Physics</u>, John Wiley (5th edition) (1976).

COUPLED NONLINEAR OSCILLATORS
J. Chandra and A.C. Scott (eds.)

SINGULARLY PERTURBED DELAY-DIFFERENTIAL EQUATIONS

Shui-Nee Chow
and
John Mallet-Paret

Mathematics Department
Michigan State University
East Lansing, MI 48824

In this paper a class of singularly perturbed delay-differential equation which, among other applications, describes an optical bistable device. In particular, the paper investigates chaotic behavior for certain parametric values. Earlier studies by others have predicted such results numerically and confirmed experimentally.

In recent years a number of interesting nonlinear differential-delay equations of the form

$$\varepsilon\dot{x}(t) = -x(t) + f(x(t-1), \lambda) \ , \ t \geq 0 \tag{1}$$

have occurred in applications, where $\varepsilon > 0$ is a small parameter, x is a scalar variable and f is a nonlinear function depending on a parameter λ. For example, the equation that describes an optically bistable device is

$$\varepsilon\dot{x}(t) = -x(t) + \lambda[1 - \sin x(t-1)]$$

In [6] Ikeda showed numerically and it was confirmed experimentally by Gibbs, Hopf, Kaplan and Shoemaker [2] that instability or chaotic behavior occurs for small ε and certain values of λ. In the study of evolutionary biology, Wazewska-Czyzewska and Lasota [7] showed that the equation that describes the production of red blood cells is

$$\varepsilon\dot{x}(t) = -x(t) + \lambda x(t-1)^8 \ e^{-x(t-1)}$$

Other examples can also be found in Glass and Mackey [3] and Hoppensteadt [5].

A primary concern is the question of how the resulting dynamical system in an infinite dimensional phase space for equation (1) is related to the one dimensional system

$$x_{n+1} = f(x_n, \lambda) \tag{2}$$

which is obtained by setting $\varepsilon = 0$ and $x_n = x(n)$. If $x(\theta) = \phi(\theta)$, $-1 \leq \theta \leq 0$, then for $\varepsilon > 0$ (1) determines a unique solution $x(t)$ for $t \geq 0$. This is found simply by integrating the equation

$$\varepsilon\dot{x}(t) = -x(t) + g(t) \ , \ 0 \leq t \leq 1$$

$$x(0) = \phi(0) \ ,$$

where $g(t) = f(x(t-1),\lambda)$, to obtain $x(t)$ for $0 \leq t \leq 1$. Successive integrations determine $x(t)$ for all $t \geq 0$. As in Hale [4], let $C = C[-1,0]$ be the Banach space of all continuous functions on the interval $[-1,0]$ with the usual sup norm. If $y(\cdot)$ is a continuous function on some interval, then let y_t denote the element in C defined by $y_t(\theta) = y(t+\theta)$, $-1 \le \theta \le 0$. Thus, given $x_0 = \phi \in C$ we uniquely obtain $x(t)$ for $t \ge 0$ and therefore $x_t \in C$ for $t \ge 0$. This may be formalized by introducing the nonlinear evolution map

$$T(t,\varepsilon) : C \to C \quad t \ge 0$$

given by $T(t,\varepsilon)\phi = x_t$. As (1) is autonomous, we have that $T(t,\varepsilon)\, T(s,\varepsilon) = T(t+s,\varepsilon)$, for all $t, s \ge 0$.

Consider now the time one map $S(\varepsilon) = T(1,\varepsilon)$. We have for any $\phi \in C$ and $-1 \le \theta \le 0$,

$$s(\varepsilon)\phi(\theta) = \phi(0)e^{-\frac{\theta+1}{\varepsilon}} + \frac{1}{\varepsilon}\int_{-1}^{\theta} e^{-\frac{\theta-s}{\varepsilon}} f(\phi(s),\lambda)\, ds \, .$$

Thus, for any $\delta > 0$ we have uniformily on $[-1+\delta,0]$,

$$S(\varepsilon)\phi(\theta) \to f(\phi(\theta),\lambda) \ , \text{ as } \varepsilon \to 0 \, .$$

In a gross sense therefore $S(\varepsilon)$ is close to the mapping f. In terms of the solution $x(t)$, we have for any $n \ge 1$ and $\delta > 0$ fixed,

$$x(n+\theta) \to f^{(n)}(x(\theta),\lambda) \ , \text{ as } \varepsilon \to 0$$

uniformly for $\theta \in [-1+\delta,0]$, where $f^{(n)} = f \circ \cdots \circ f$ is the nth iterate of f. In this way, (2) may be considered as a singular limit of (1). In the following, we will describe some results relating equations (1) and (2).

Suppose for some λ, a is an exponentially stable fixed point of f under iteration, that is

$$f(a\lambda) = a \ , \ |f_x(a,\lambda)| < 1 \, . \tag{3}$$

By considering the characteristic equation of the linearized differential-delay equation we have the following.

Proposition 1. Suppose (3) holds. Then for any $\varepsilon > 0$, $x(t) \equiv a$ is a uniformly asymptotically stable solution of (1).

Suppose the fixed point a of f loses its stability as λ passes through λ_1. We expect the stability of the constant solution $x(t) \equiv a$ will also change as λ passes through λ_1. In fact, we have the following.

Theorem 2. Suppose that $f(a,\lambda) = a$ and $f_x(a,\lambda) = -1 - \alpha(\lambda-\lambda_1) + O(|\lambda-\lambda_1|^2)$ for λ near λ_1 where $\alpha > 0$ is a constant. Then the constant solution $x(t) \equiv a$ of (1) loses its stability along a curve (see Fig. 1) Γ:

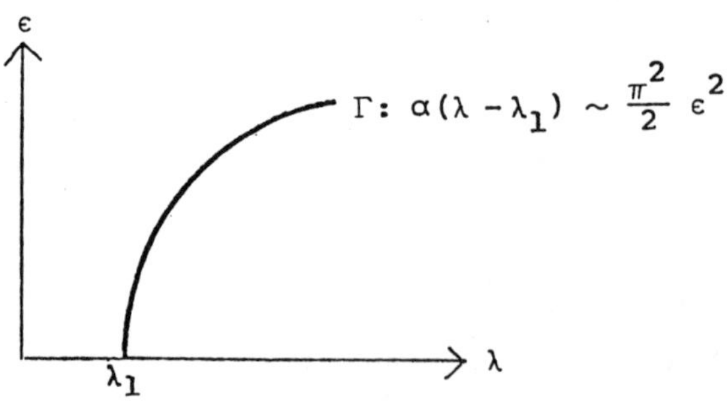

Figure 1

$$\alpha(\lambda-\lambda_1) = \frac{\pi^2}{2}\varepsilon^2 + 0(\varepsilon^3) \ , \ (\lambda,\varepsilon) \text{ near } (\lambda_1,0) \ , \tag{4}$$

on which a pair of purely imaginary eigenvalues near $\pm\, i\pi$ of the characteristic equation of the linearization of (1) at $x(t) \equiv a$ occurs. Indeed, $x(t) \equiv a$ is exponentially stable for parameters to the left of Γ, and unstable to the right, in a neighborhood of $(\lambda,\varepsilon) = (\lambda_1,0)$.

It can be shown by the methods in [1] that a Hopf bifurcation occurs along the curve Γ (4). Furthermore, numerical studies indicate that the bifurcations are often supercritical for classes of equations arising in applications. The bifurcating stable periodic solutions may be approximated by $a + c\sin(\pi t)$, where $c > 0$ is the amplitude, for $(\lambda-\lambda_1,\varepsilon)$ small and near the curve (4). We note that the bifurcating periodic solutions have minimal periods approximately equal to 2. On the other hand, for the discrete mapping f, λ_1 is the value of the parameter at which the fixed point a loses its stability and bifurcates into a pair of points of period 2. Since the time one map $S(\varepsilon)$ is close to the mapping f, this indicates the existence of a "square" wave periodic solution with period approximately 2 of equation (1) provided (ε,λ) is close to $(0,\lambda_1)$ and $\lambda > \lambda_1$. In fact, numerical studies strongly indicate the existence of such a square wave. Furthermore, these square waves turn into sine waves as ε increases, with λ fixed.

To be precise, fix $\lambda > \lambda_1$, near λ_1, and let $\varepsilon^*(\lambda)$ denote the value of ε such that $(\lambda,\varepsilon) \in \Gamma$. Now vary ε with λ fixed. For ε near $\varepsilon^*(\lambda)$ (typically on one side of $\varepsilon^*(\lambda)$) the Hopf bifurcation solutions exist. They resemble sine waves. On the other hand for ε near 0 (how near depending on λ) we anticipate solutions of (1) close to square waves of period 2 to exist. (See Fig. 2) Furthermore,

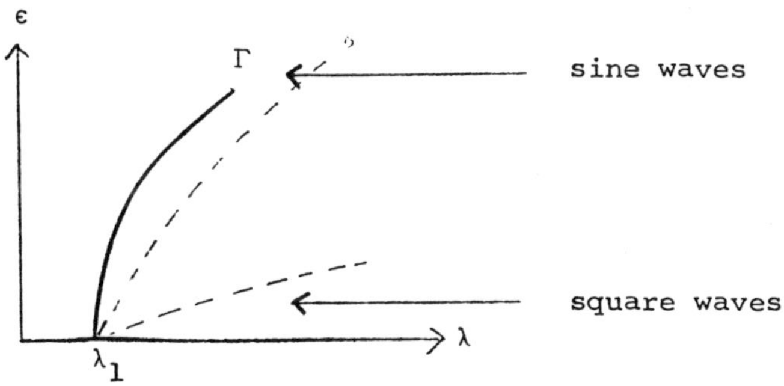

Figure 2

equation (1) parameterized by ε should possess a connected set of periodic solutions near a, with period near 2, and parameter ε ranging from 0 to $\varepsilon^*(\lambda)$, containing both the Hopf solutions and the square wave solutions.

We now consider the square waves in greater detail. Here, we will consider a formal approach. Suppose for a particular λ the mapping f has a pair of points of period 2 (for example arising from a bifurcation from a, as mentioned, for λ near λ_1):

$$f(a,\lambda) = a_2, \ f(a_2,\lambda) = a_1 > a_2 \ .$$

We wish to obtain equations describing the boundary layer of a square wave $x(t)$ i.e., the ranges of values of t near which $x(t)$ suffers an abrupt jump. We expect $x(t)$ to be alternately near a_1 and a_2 for time intervals of length 1 +

$O(\varepsilon)$ and to pass from a_i $(i = 1,2)$ to $f(a_i)$ on intervals of length $O(\varepsilon)$; in particular the period of $x(t)$ would then be $2 + O(\varepsilon)$. (See Fig. 3.)

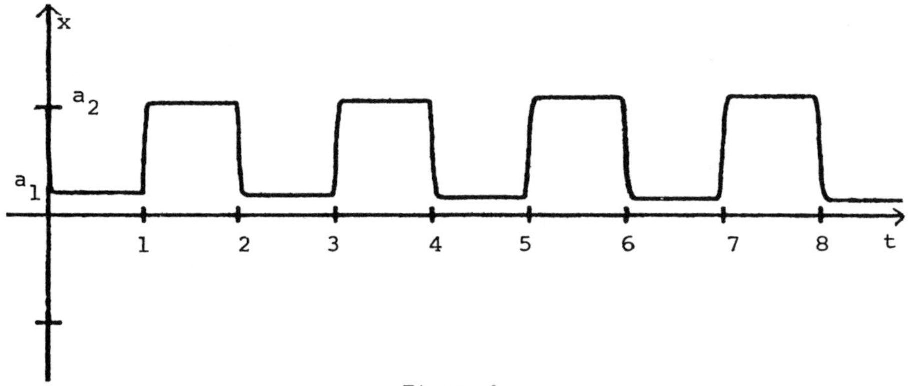

Figure 3

This suggests the following change of variables:

$$\varepsilon\tau = -t$$

$$y_1(\tau) = x(t) \tag{5}$$

$$y_2(\tau) = x(t + 1 + r\varepsilon)$$

where the period of $x(t)$ is assumed to $2 + 2r\varepsilon + O(\varepsilon)$, for some constant r. Using the periodicity of $x(t)$ and formally taking limits as $\varepsilon \to 0$, we obtain differential equations

$$\begin{aligned} y_1'(\tau) &= y_1(\tau) - f(y_2(\tau-r)\ ,\ \lambda)\ , \\ y_2'(\tau) &= y_2(\tau) - f(y_1(\tau-r)\ ,\ \lambda)\ , \end{aligned} \tag{6}$$

with boundary conditions

$$\begin{aligned} (y_1(\tau),y_2(\tau)) &\to (a_2,a_1) \quad \text{as} \quad \tau \to -\infty \\ (y_1(\tau),y_2(\tau)) &\to (a_1,a_2) \quad \text{as} \quad \tau \to \infty \end{aligned} \tag{7}$$

(We note that by rescaling time $\tau \to r\tau$, the differential equations (6) are converted to ones with delay 1 and a parameter r:

$$y_1'(\tau) = ry_1(\tau) - rf(y_2(\tau-1),\lambda)\ ,$$

$$y_2'(\tau) = ry_2(\tau) - rf(y_1(\tau-1),\lambda)\ .$$

Such a system is thus covered by the general theory of Hale referred to above.)

Observe that (a_2,a_1) and (a_1,a_2) are both critical points of Eq. (6). Our concern is thus the existence of a heteroclinic orbit for some value of r, that is, a trajectory joining these critical points at times $\tau = \pm\infty$. The delay parameter r is unknown a priori, and must be determined as part of the solution. By normalizing the fixed point of f at a = 0 and parameter so that $\lambda_1 = 1$ and $\alpha = 1$, we have the following theorem.

Theorem 3 *Suppose that near* $x = 0$

$$f(x,\lambda) = -\lambda x + px^2 + qx^3 + 0(x^4)$$

where p and q are constants and $p^2 + q > 0$. *Then for* $\lambda - 1 > 0$ *sufficient small, the problem (6), (7) has a solution* (y_1,y_2) *with* $r = 1 - 4/15\ (\lambda-1) + 0(|\lambda-1|$. *Furthermore,* $y_1(\tau)$ *and* $y_2(\tau)$ *are monotone functions of* τ.

Outline of Proof. For simplicity assume the symmetry condition $f(-x,\lambda) = -f(x,\lambda)$, $p = 0$, $q > 0$, and look for solutions such that $y_1(\tau) = -y_2(\tau)$. (This corresponds to a solution $x(t)$ of (1) satisfying $x(t+1+r\varepsilon+0(\varepsilon)) = -x(t)$.) Then the problem (6), (7) is reduced to

$$\begin{aligned} &y'(\tau) = y(\tau) + f(y(\tau-r),\lambda) \\ &y(\tau) \to \pm\ a_1 \quad \text{as} \quad \tau \to \pm\ \infty\ . \end{aligned} \tag{8}$$

A rescaling sets $q = 1$, so without loss $f(x,\lambda) = -\lambda x + x^3 + 0(x^4)$. Thus, $a_1 \sim\sqrt{\lambda-1}$ for $\lambda > 1$. Motivated by the structure of the stable and unstable manifolds of the constant solutions $y(\tau) = \pm\ a_1$, we assume that $r(\lambda) =$

$1 + r_1\mu + r_2\mu^2 + \cdots$ where $\mu = \sqrt{\lambda-1}$. Let $s = r^{-1}\mu\tau$ be a new time variable and $y(\tau) = \mu w(s)$. We obtain the following form (8):

$$\begin{aligned} &\mu\,\frac{dw}{ds} = r(\lambda)[w(s) - (1+\mu^2)\ w(s-\mu) + \mu^2 w(s-\mu)^3 + 0(\mu^3)] \\ &w(s) \to \pm\ 1 + 0(\mu) \quad \text{as} \quad s \to \pm\ \infty\ . \end{aligned} \tag{9}$$

Expand the above equation in powers of μ, including an expansion in the delay: $w(s-\mu) = w(s) - \mu\ dw/ds + 1/2\ \mu^2\ d^2w/ds^2 + \cdots$. Also write $w(s) = w_0(s) + \mu w_1(s) + \mu^2 w_2(s) + \cdots$. Now equate coefficients of μ^k in this formal expansion. For $k = 0$ and 1 the coefficients agree. The coefficients of μ^2 yield the following:

$$\begin{aligned} &\frac{1}{2}\,\frac{d^2w_0}{ds^2} + r_1\,\frac{dw_0}{ds} + w_0 - w_0^3 = 0 \\ &w_0(s) \to \pm\ 1 \quad \text{as} \quad s \to \pm\ \infty \end{aligned} \tag{10}$$

This damped Hamiltonian equation has the unique solution $w_0(s) = \tanh s$ and $r_1 = 0$. We obtain $r_2 = -\ 4/15$ by considering the coefficients of μ^3 and $w_1(s)$. This method gives an approximate solution to problem (9). An existence proof of problem (9) is based upon the knowledge of the approximate solution, and careful estimates of the resolvent of the delay-differential operator obtained by the Fourier transform.

Theorem 3 is a perturbation result. For more global results, we need more hypotheses on f. We state the following results for a nonlinearity $f(x)$ in which the parameter λ is absent.

Assume that $f(-x) = -f(x)$, $f(a) = -a$ for some $a > 0$, and that $f(x) > -x$ on $(-a,0)$. See Figs. 4 and 5.

Theorem 4. *If in addition to the above,* $f'(x) < 0$ *on* $[-a,a]$ *with* $f'(0) < -1$, *then there exists* $r > 0$ *such that (8) (with* λ *absent and* $a = a_1$*) has a monotone solution.* (See Fig. 4.) *Moreover this r is unique, as is the solution.*

Theorem 5. *Assume in addition that* $0 < f'(a) < 1$, $f'(0) < -1$, *that f is monotone increasing on* $(-\infty,-\xi)$ *and monotone decreasing on* $(-\xi,0]$ *for some*

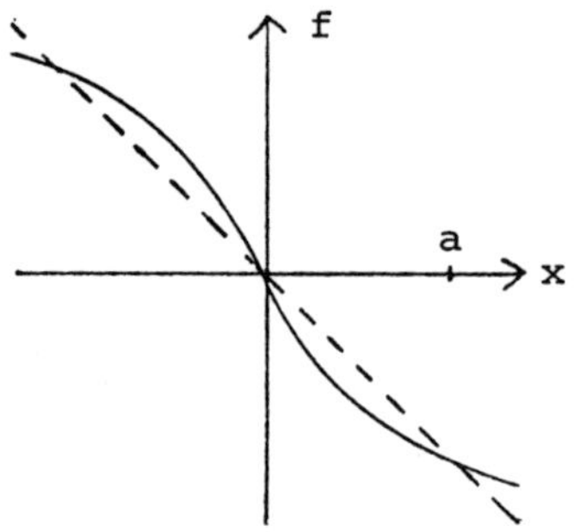

Figure 4

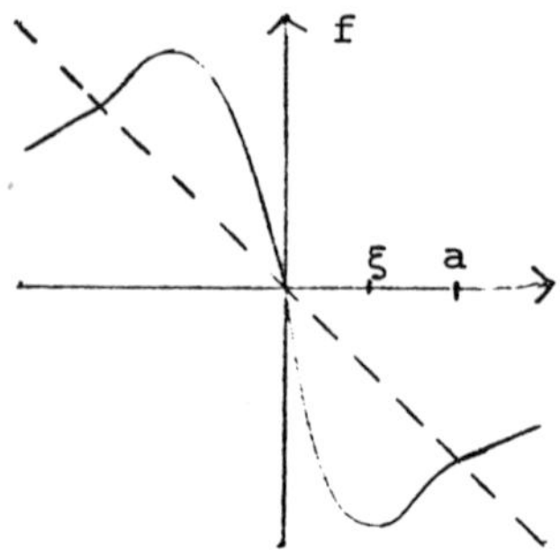

Figure 5

$0 < \xi < 1$. (See Fig. 5.) Assume further that the points $\pm$ a attract all points x [-a,a] - {0} upon iteration of f. Then (8) has no monotone solution. Moreover there exists $r > 0$ such that (8) has a solution which oscillates about a as $\tau \to \infty$ (that is, sup $y(\tau) > a$ and $y(\tau) = a$ infinitely often on any interval $[\theta_0, \infty)$.)

REFERENCES

[1] S. N. Chow and J. Mallet-Paret, Integral averaging and bifurcation. J. Diff. Eqns., 26 (1977), 112-159.

[2] H. M. Gibbs, F. A. Hopf, D. L. Kaplan and R. L. Shoemaker, Observation of chaos in optical bistability. Phys. Rev. Lett., 46 (1981), 474-477.

[3] L. Glass and M. Mackey, Oscillation and chaos in physiological control systems. Science, 197 (1977), 287-289.

[4] J. K. Hale, Theory of Functional Differential Equations. Springer-Verlag, Berlin, 1977.

[5] F. C. Hoppensteadt, Mathematical theories of population: demographics, genetics and epidemics. S.I.A.M. Publ. Phila., 1976.

[6] K. Ikeda, Multiple-valued stationary state and its instability of the transmitted light by a ring cavity systems. Opt. Commun. 30, 257 (1979).

[7] M. Wazewska-Czyzewska and A. Lasota, Mathematical models of the red cell system (Polish). Mate. Stosowana, 6 (1976), 25-40.

COUPLED NONLINEAR OSCILLATORS
J. Chandra and A.C. Scott (eds.)

CHAOS IN SIMPLE LASER SYSTEMS

Robert Graham

Fachbereich Physik
Universität Essen, W. Germany

We discuss the appearance of chaos in the nonlinear dynamics of various laser systems.

1. INTRODUCTION

The theory of lasers due to Haken [1] and Lamb [2] reduces the dynamics to be studied to that of one or several coupled nonlinear oscillators, classical or quantized. Indeed, phenomena familiar from nonlinear oscillators like limit cycles, frequency pulling, phase- and frequency locking between various modes, combination tones are all commonly observed in lasers.

These phenomena were all appreciated and discussed when the Haken-Lamb theory was first formulated. However, an additional feature of nonlinear oscillators has not yet received much study in lasers: the ability of nonlinear oscillators for chaotic motions. The notion "chaos" is used here in its technical sense to indicate "sensitive dependence on initial conditions", associated with erratic motion on the energy hypersurface in conservative systems [3]. The present paper is devoted to the discussion of various cases, where chaos appears in simple laser system.

2. SINGLE MODE INTERACTING WITH ATOMS

We start our discussion with some remarks on the conservative dynamics of a single mode in dipole coupling with a system of identical two level atoms. The Hamiltonian is

$$H = \varepsilon S_z + \hbar\omega b^+ b + g(b^+ + b)(S_+ + S_-) \tag{2.1}$$

with the dipole coupling constant g, mode frequency ω, and level spacing $\hbar\varepsilon$. We have described the two level atoms by Pauli matrices σ_μ and have introduced the formal spin operators

$$\vec{S} = \frac{\hbar}{2}\sum_\mu \vec{\sigma}_\mu \; ; \; S_\pm = S_x \pm iS_y \tag{2.2}$$

The hermitian operator $S_z/\hbar$ represents the observable "one half of the population inversion of the two atomic levels" in the system. The application of the non-hermitian operators $S_\pm$ to any eigenstate of S_z leads to a new eigenstate of S_z whose eigenvalue differs by $\pm\hbar$, except for the fully excited eigenstate of S_z, which is annihilated by S_+ and the fully de-excited eigenstate of S_z, which is annihilated by S_-.

The mode is quantized and described by Bose operators b, b^+, satisfying the commutation relations $[b, b^+] = 1$; $[b,\vec{S}] = 0$. The hermitian operator b^+b represents the observable "number of quanta in the mode". The application of b^+ leads to a new eigenstate, with one quantum more or less, except for the ground state of b^+b, which is annihilated by b.

The first two terms of the Hamiltonian (2.1) represent the energy of the uncoupled atoms and single mode, respectively. The last term of (2.1) describes the exchange of energy between the atoms and the mode. It contains the "resonant terms" $g(b^+S_-+bS_+)$ which (for $\varepsilon \approx \omega$) describe a real exchange of energy quanta $\hbar\varepsilon$ between atoms and mode, and it also contains the "anti-resonant" terms $g(b^+S_++b^+S_-)$ which describe a virtual exchange of energy between atoms and mode and violate energy conservation by $|\Delta E| = \hbar(\omega+\varepsilon)$. Due to the uncertainty principle, virtual transitions are allowed, but are restricted to short time intervals

$$\Delta t \cong \hbar/|\Delta E| \;.$$

A complete set of commuting observables of this system is e.g. $b + b^+$, S^2, S_z, i.e., the system has f = 3 degrees of freedom. The Hamiltonian (2.1) has the commuting constants of the motion H, S^2. Other independent constants of the motion of (2.1) are not known. Since their number is less than f, the dynamics generated by (2.1) may be expected to display chaos [3,4] in some regions of phase space. A study of (2.1) from this point of view has not yet been carried out. It is therefore also possible that (2.1) still contains a hidden constant of the motion, in which case it would be exactly integrable, and chaos could not appear. Chaos in quantum systems has been discussed in [4].

An approximate version of (2.1) is obtained by introducing the rotating wave approximation. This approximation amounts to neglecting the "anti-resonant terms" in Eq. (2.1). For optical frequencies this is quantitatively a very good approximation, which breaks down only on very short time scales $\Delta t \cong 1/2\varepsilon$. However, for subtle qualitative questions like the appearance of chaos, this approximation can still have drastic consequences, as will now be shown. In the rotating wave approximation the Hamiltonian (2.1) reduces to

$$H = \varepsilon\, S_z + \hbar\omega b^+b + g(b^+S_-+bS_+) \tag{2.3}$$

The independent constants of the motion commuting with each other are now H, S^2, $S_z + b^+b$.

Since their number now equals f, the system (2.3) is exactly integrable [3,4], and has, indeed, been solved exactly [5]. The exact quantum dynamics turns out to be still amazingly complicated [6], but it is quasi-periodic and not chaotic due to the existence of 3 independent constants of the motion.

3. SINGLE MODE LASER AND THE LORENZ MODEL

The equations of motion of a homogeneously broadened single mode laser in rotating wave and semiclassical approximation are [1]

$$\begin{aligned} \dot\beta &= -(i\varepsilon+\mathcal{H})\beta - \frac{ig}{\hbar}S_- \\ \dot S_- &= -(i\varepsilon+\gamma_1)S_- + 2ig\beta S_Z \\ \dot S_Z &= \gamma_{11}(S_o-S_Z) + ig(\beta^*S_--\beta S_+) \end{aligned} \tag{3.1}$$

We have used the notation of Eq. (2.1) but replaced b, b^+ by c-numbers β,β^* and also treat the spin classically. In addition we introduced the damping constants , γ_1, γ_{11} and the pumping rate $\gamma_{11}S_o$ and assumed perfect tuning $\omega = \varepsilon$. The Maxwell Bloch equations (3.1) are closely related to the equations of the Lorenz model [7], as was first pointed out by Haken [8]. The Lorenz model first appeared as an approximation to the equations of thermal convection in a fluid

[9], but is now known to occur also in many other cases [10]. It is obtained from Eq. (3.1) for long times, where the phases of β and S_- are locked

$$\beta = Ee^{-i\rho-i\varepsilon t} \ , \ S_- = iPe^{-i\rho-i\varepsilon t} \tag{3.2}$$

with ρ independent of time, and real E, P. The Lorenz model is known [7] to exhibit a transition to chaos under certain conditions. For the laser, these conditions are only satisfied if

$$H > \gamma_1 + \gamma_{11} \tag{3.3}$$

which is physically unrealistic, since in lasers H/γ_1 is usually made very small. On the other hand, Eq. (3.3) may be satisfied in situations designed to create superfluorescence [11] where H/γ_1 is made as large as possible in order to avoid a quenching of the radiating coherent atomic dipole moment by a feedback from the generated electromagnetic field which is 180° out of phase [12].

If Eq. (3.3) is satisfied, chaos always appears in the dynamics of Eq. (3.1) for

$$S_o > \frac{\hbar}{2}\,\frac{H^2(H+\gamma_{11}+3\gamma_1)}{g^2(H-\gamma_1-\gamma_{11})} \tag{3.4}$$

and shows up as a random pulsing of the amplitude of the single mode [8]. For a range of values of S_o below (3.4) the system (3.1) is bistable and a stable fixed point (corresponding to cw single mode laser action) coexists with a stable strange attractor (corresponding to random single mode pulsing)[13]. The co-existence of a stable fixed point and a stable or nearly stable (in the sense that it traps the system for a long time) strange attractor may still be possible if (3.3) is not satisfied and might be one explanation for the random single mode spiking sometimes observed in lasers.

4. MODE LOCKED LASERS AND THE LORENZ MODEL

Above a certain threshold of the pumping rate single mode lasers become unstable against the formation of pulse trains, consisting of many modes whose phases are locked by their nonlinear interaction [14,15]. This instability occurs even if $H < \gamma_1 + \gamma_{11}$. If we consider pulse trains in an infinitely extended laser medium which are time independent in a co-moving frame, we can again make contact with the Lorenz model [16]. The condition (3.3) is, in this case, replaced by

$$C < V < C\,\frac{\gamma_1+\gamma_{11}}{\gamma_1+\gamma_{11}-H} \tag{4.1}$$

where V is the phase velocity of the random pulse train, appearing at the instability. Since V is not restricted by external conditions, condition (4.1) can always be satisfied for $H < \gamma_1 + \gamma_{11}$. A chaotic pulse train appears first with phase velocity

$$V = C \quad 1 + H(\gamma_1+\gamma_{11}- \ +2\sqrt{(\gamma_1+\gamma_{11})(2\gamma_1+\gamma_{11})})^{-1} \tag{4.2}$$

if S_o satisfies

$$S_o \geq \frac{\hbar}{2}\,\frac{H}{g^2}\,(3\gamma_{11}+5\gamma_1+\sqrt{8(\gamma_{11}+\gamma_1)(\gamma_{11}+2\gamma_1)}) \tag{4.3}$$

It is very interesting to note that a phase velocity V > C is in this case associated with a <u>random</u> pulse train, since one is used to allow for this possibility for periodic pulse trains only. However, the relativity principle is not in jeopardy. Since the pulse train corresponding to the strange attractor is random, any finite subsection of the train does not allows one to determine from which part of the pulse train it has been taken, and thus cannot be used to carry information or energy.

If ring lasers of finite length are considered instead of infinitely long lasers, periodic boundary conditions must be imposed on the pulse trains [15]. In this case the Lorenz model is still applicable [16] to pulse trains traveling distortionless, but only its periodic solutions are now relevant, which would be unstable in the absence of periodic boundary conditions. They describe periodic pulse trains with $V > C$. The numerical work in [15] shows that these periodic solutions have a finite basin of attraction. Such pulse trains are observed in mode locked lasers.

In addition to these periodic solutions in ring lasers, chaotic solutions were also found in numerical work by M. Mayr et. al. [17], starting from different initial conditions. These solutions are time dependent also in the co-moving frame. Chaotic solutions were also found in ring lasers after dropping the rotating wave approximation [18].

5. SINGLE MODE LASERS DRIVEN BY EXTERNAL FIELDS

Chaotic dynamics has been successfully forced on hydrodynamics systems near instabilities by perturbing them periodically [19]. This mechanism can also be used to force chaotic dynamics on single mode oscillators. Two examples have recently been discussed in the literature [20,21].

The first example [20] is a single mode laser above threshold driven by an amplitude modulated coherent external field. The equation of motion of the mode amplitude takes the form

$$\frac{d\beta}{d\tau} = -(i\Omega+1)\beta + \frac{2g^2 S_o \beta}{\hbar H \gamma_1 \left(1+\frac{4g^2}{\gamma_{11}\gamma_1}|\beta|^2\right)} + A(\tau) \tag{5.1}$$

It is obtained from (3.1) for $H \ll \gamma_1, \gamma_{11}$ after adding a driving field with frequency $\omega = \varepsilon - H\Omega$ and modulated amplitude $A(\tau)$, splitting off the main frequency ω and eliminating the atomic degrees of freedom in adiabatic approximation. Time has been scaled in (5.1) by putting $\tau = Ht$.

The effect of the external field A on the laser is first of all to increase the laser threshold [20]. A similar phenomenon is now known to occur also in other systems, like the Bénard instability [22]. Above threshold the laser amplitude moves on a limit cycle in the complex β-plane because of the presence of finite detuning $\Omega \neq 0$. This limit cycle shows up in the laser spectrum as a number of equidistant sharp satellite lines in the neighbourhood of the fundamental frequency ω.

If the external field is periodically amplitude modulated with a frequency $H\Omega'$

$$A(\tau) = a + a' \cos\Omega'\tau \tag{5.2}$$

chaos can be induced in the system. Numerical solutions of Eqs. (5.1), (5.2) show that with increasing a' the limit cycle is first locked to a rational part of Ω' and then looses its stability to a state with continuous spectral side bands in the H-vicinity of the main frequency ω, and sensitive dependence on initial conditions [20].

Another example [21] is relevant to cases where a laser mode is driven by an external coherent cw source at a nearby frequency and the atomic inversion S_o seen by this mode is periodically modulated. This situation may occur very naturally in multi-mode lasers above threshold where fields at combination tones are produced by the nonlinearities and drive modes at nearby frequencies, while the net inversion felt by a given mode is time dependent due to population number pulsations (cf. [1,2]).

The sequence of states obtained in this model with increasing modulation depth of the pumping rate is again a quasi-periodic state consisting of a perturbed limit cycle, a locked state where the perturbed limit cycle closes on itself after three revolutions, and a chaotic state. The mechanism for the onset of chaos was investigated in some detail for this example [21]. It turned out that chaos appears here via intermittency by a general mechanism first described by Manneville and Pomeau [23]. The same onset mechanism is likely to be at work also in the preceeding example, as the bifurcation schemes and the spectra in the chaotic states are very similar for both cases. If the modulation depth becomes too large, the chaotic state is returned to a periodic state of ever shorter period via an inverted Feigenbaum sequence [24] of period undoubling bifurcations.

6. "STATISTICAL LIMIT CYCLES" IN MULTI-MODE LASERS

In the preceeding examples laser systems have been discussed which fail to settle down to a time independent or quasi-periodic state due to the appearance of a strange attractor in their phase space. There is yet another way, recently discussed by Busse and Heikes [25] and Kramer [26], in which a system may erratically wander in phase space. However, different from the cases considered before, the basic mechanism here requires the presence of some additional fluctuations in the system, even if these are allowed to be small. The mechanism is based on a cyclic instability between several fixed points in phase space. E.g. if the system is in state A a small fluctuation may be sufficient to bring it to state B (but not back), from there to C etc. and finally back to A. Busse, who has found an example of such a behaviour in a rotating Bénard cell, has coined the term "statistical limit cycle" for this sort of cyclic but nonperiodic dynamics.

Statistical limit cycles may well appear in multi-mode lasers, e.g. in a 3 mode laser, if it happens that mode 1 preferentially drives mode 2, which preferentially drives mode 3, which drives mode 1 again. In such a laser one will observe the cyclic but statistical appearance and disappearance of the three modes involved, triggered by the spontaneous emission into the three modes. The equations governing this phenomenon for 3 intensity coupled modes are given by the Haken-Lamb theory [1,2] as

$$\dot{I}_1 = 2\, I_1(a_1-b_1I_1-b_{12}I_2-b_{13}I_3) + \sqrt{q_1I_1}\; \xi_1(t)$$

$$\dot{I}_2 = 2\, I_2(a_2-b_2I_2-b_{21}I_1-b_{23}I_3) + \sqrt{q_2I_2}\; \xi_2(t)$$

$$\dot{I}_3 = 3\, I_3(a_3-b_3I_3-b_{31}I_1-b_{32}I_2) + \sqrt{q_3I_3}\; \xi_3(t)$$

$$\langle \xi_i(t)\; \xi_j(0)\rangle = \delta_{ij}\; \delta(t)$$

Where I_i are the intensities involved, $\xi_i(t)$ are stochastically independent Gaussian white noise sources describing spontaneous emission, which turns Eqs. (6.1) into stochastic differential equations in the sense of Itö.

The parameters a_i are positive if all three modes are above threshold. The parameters b_i, b_{ij} describe selfsaturation and cross-saturation effects, respectively.

Equations (6.1), apart from the noise sources, are identical to equations solved numerically in [25,26]. A statistical limit cycle appears in Eq. (6.1) if all $a_i > 0$, if cross saturation dominates over self saturation

$$b_{ij} > \frac{a_i}{a_j} b_j \tag{6.2}$$

and if the cross saturation coefficients define a sense of revolution by

$$\left(\frac{a_2}{a_1}\right)^2 b_{12} > b_{21} \ ; \ \left(\frac{a_3}{a_2}\right)^2 b_{23} > b_{32} \ ; \ \left(\frac{a_1}{a_3}\right)^2 b_{31} > b_{13} \ . \tag{6.3}$$

The three states whose neighbourhoods in phase space are visited during one cycle are (in an obvious vector notation for the three intensities (I_1, I_2, I_3))

$$\left(\frac{a_1}{b_1}, 0, 0\right) \ ; \ \left(0, \frac{a_2}{b_2}, 0\right) ; \ \left(0, 0, \frac{a_3}{b_3}\right). \tag{6.4}$$

The transitions between these states occur only assisted by fluctuations. The average period of the cycle therefore strongly depends on the noise intensities q_i and approaches infinity if the noise vanishes.

7. CONCLUSIONS

Lasers are nonlinear oscillators par excellence and display all the phenomena typical of such systems. The appearance of chaos is no exception, as was shown in this paper for a few simple models. Chaos should also appear in many realistic laser systems, in particular in multi-mode lasers, where the nonlinear interaction between the modes should typically lead to the appearance of a strange attractor [27]. It is not yet obvious what would be the most favourable experimental set up for investigating chaos in lasers. But perhaps something can be gleaned from the very successful experimental work on similar phenomena in hydrodynamic systems [28-31]. One would then be led to prefer the simplest and most controlled systems available, e.g. He-Ne single mode lasers, and try to induce chaos by enlarging their phase space in a carefully controlled way, possibly by one of the mechanisms described in section 5 of this paper.

ACKNOWLEDGMENT

The talk on which this paper is based was prepared during the author's stay at the Institute for Theoretical Physics in Santa Barbara. I wish to thank this institution for its hospitality, its members for useful discussions - in particular Pierre Hohenberg, Manfred Lücke, Crispin Gardiner, and Fritz Haake - and the Deutsche Forschungsgemeinschaft and the National Science Foundation for financial support.

REFERENCES

1. H. Haken, H. Sauermann, Z. Physik 173, 261 (1963); 176, 47 (1963); H. Haken, Laser Theory, Encyclopedia of Physics 25/2c (1970).

2. W. E. Lamb, Jr., Phys. Rev. 134A, 1429 (1964); M. Sargent III, M. O. Scully, W. E. Lamb, Jr., Laser Physics, Addison-Wesley, London 1974.

3. R. H. G. Hellemann, in "Fundamental Problems in Statistical Mechanics," vol. 5, ed. E. G. D. Cohen, North Holland, Amsterdam 1980.

4. M. V. Berry, in "Topics in Nonlinear Dynamics," ed. S. Jorna, Am. Inst. Phys. Conf. Proc. Series 46, 16 (1978).

5. E. T. Jaynes, F. W. Cummings, Proc. I.E.E.E. 51, 89 (1963).

6. cf. J. H. Eberly's contribution to this workshop and references given there.

7. E. N. Lorenz, J. Atmos. Sci. 20 130 (1963).

8. H. Haken, Phys. Lett. 53A, 77 (1975).

9. B. Saltzman, J. Atmos. Sci. 19, 329 (1962).

10. C. J. Marzec, E. A. Spiegel, SIAM J. Appl. Math. 38, 403 (1980).

11. F. Haake, R. Glauber, Phys. Rev. A5, 1457 (1972).

12. R. Graham, in "Progress in Optics," vol. 12, ed. E. Wolf, North Holland, Amsterdam 1974, p. 246.

13. J. Kaplan, J. A. Yorke, Conference on Bifurcation Theory and Applications in Scientific Disciplines, New York Academy of Sciences, 1977.

14. R. Graham, H. Haken, Z. Physik 213, 420 (1968).

15. H. Risken, H. Nummedal, J. Appl. Phys. 39, 4662 (1968).

16. R. Graham, Phys. Lett. 58A, 440 (1976).

17. M. Mayr, H. Risken, H. D. Vollmer, Optics Comm. 36, 480 (1981).

18. H. Haken, in "Dynamics of Synergetic Systems," ed. H. Haken, Springer, Berlin 1980.

19. J. P. Gollub, S. V. Benson, Phys. Rev. Lett. 41, 948 (1978).

20. T. Yamada, R. Graham, Phys. Rev. Lett. 45, 1322 (1980).

21. H. J. Scholz, T. Yamada, H. Brand, R. Graham, Phys. Lett. 82A, 321 (1981).

22. M. Lücke, private communication.

23. P. Manneville, Y. Pomeau, Phys. Lett. 75A, 1 (1979).

24. M. Feigenbaum, J. Stat. Phys. 19, 25 (1978); 21, 669 (1979).

25. F. H. Busse, K. E. Heikes, Science 208, 173 (1980).

26. L. Kramer, Z. Physik B41, 357 (1981).

27. D. Ruelle and F. Takens, Comm. Math. Phys. 20, 167 (1971).

28. J. P. Gollub and S. V. Benson, J. Fluid Mech. 100, 449 (1980).

29. M. Dubois, P. Bergé, J. Fluid Mech. 85, 641 (1978).

30. G. Ahlers, R. P. Behringer, Progr. Theor. Phys. Suppl. 64, 186 (1979).

31. J. Maurer, A. Libchaber, J. Physique Lett. 40, L419 (1979).

COUPLED NONLINEAR OSCILLATORS
J. Chandra and A.C. Scott (eds.)

TRUNCATED NAVIER-STOKES EQUATIONS ON A TWO-DIMENSIONAL TORUS

Valter Franceschini

Center for Nonlinear Studies
Los Alamos National Laboratory
Los Alamos, New Mexico 87545

The behavior of complex truncated Navier-Stokes equations is investigated. A significant role in the phenomenology of these equations is played by a set of infinitely many invariant N-dimensional hyperplanes (subspaces of the 2N-dimensional phase space) which are symmetrically placed by virtue of a symmetry group. There is numerical evidence that, at least for Reynolds numbers below a certain critical threshold, every random initial point is captured first by one of these hyperplanes, then by an attractor on it.

Particular bifurcations can take place in the 2N-phase space. A fixed point can bifurcate into a closed curve consisting entirely of fixed points; in addition, a closed curve of fixed points can bifurcate into a torus covered entirely by either fixed points or periodic orbits.

As long as the invariant hyperplanes are attracting, a complete correspondence is found between the phenomenologies of some complex models and their previously studied real versions.

1) INTRODUCTION

A problem of great interest to physicists and mathematicians is the development of a correct approach for the study of the Navier-Stokes equations for an incompressible fluid. In spite of several works, theoretical and otherwise, which have provided insight into the solution of the problem, there is still much to be learned and many other numerical investigations appear necessary to throw light on the problem.

One approach to the problem, conceived by Gallavotti, is based on the idea that the understanding of the physical phenomenon can be reached by numerically investigating suitable truncations of the two-dimensional Navier-Stokes equations on the torus, obtained by considering only a few components (modes) in the Fourier-expansion. Such a study, starting from a consideration of the minimal nontrivial truncation and proceeding to investigation of the consequences of adding or changing modes, can lead to a good approximation of the solution when larger models are examined.

The models studied up to now from this perspective [1-7], ranging from a minimal 5-mode model up to a 9-mode one, clearly show that the phenomenology strongly depends on those modes used for the truncation. This fact, certainly not completely unexpected, means that still larger models must be considered to obtain a qualitative behavior unaffected by addition or change of modes. It is reasonable to hope that such a goal can be achieved before the model becomes numerically untractable.

The truncated Navier-Stokes equations have proven to be interesting also in themselves as mathematical models of nonlinear differential equations exhibiting chaotic behavior. In [2] such a model first shows the very important role of period-doubling bifurcations of periodic orbits as a pathway to turbulence in multidimensional dissipative systems, and exhibits a significant compatibility with Feigenbaum's theory of universality [8,9]. In [6] an original road to chaos is found through successive bifurcations of tori. In any case, every model exhibits a very rich phenomenology, often with new and interesting features.

For reasons of simplicity and economy, these models were obtained by considering a particular solution in which each component is either real or purely imaginary, which we shall call a "real" solution. In this paper we shall consider the most general solution, for which each component is complex. By investigating this complex solution, we shall better understand the truncated Navier-Stokes equations and more deeply enter into the behavior of the already known models. Furthermore, we shall review some bifurcations, met in studying the "real" systems, from a new point of view, which leads to a number of remarkable conclusions.

2) TRUNCATED NAVIER-STOKES EQUATIONS

Consider the Navier-Stokes equations for an incompressible fluid on the torus $T^2 = [0,2\pi] \times [0,2\pi]$:

$$\begin{cases} \dfrac{\partial \underline{u}}{\partial t} + (\underline{u} \cdot \nabla)\underline{u} = -\nabla p + \underline{f} + \nu \Delta \underline{u} \\ \operatorname{div} \underline{u} = 0 \\ \displaystyle\int_{T^2} \underline{u} d\underline{x} = 0 \\ \underline{u}|_{t=0} = \underline{u}_o \end{cases} \tag{2.1}$$

where $\underline{u}$ is the velocity field, p is the pressure, ν is the viscosity and $\underline{f}$ a periodic volume force. Expand $\underline{u}$ in Fourier-series

$$\underline{u}(\underline{x}) = \sum_{\underline{k}\neq 0} \gamma_{\underline{k}}\, e^{i\underline{k}\cdot\underline{x}} \frac{\underline{k}^{\perp}}{|\underline{k}|}, \tag{2.2}$$

where $\underline{k} = (h_1,h_2)$ is a wave vector with integer components, $\underline{k}^{\perp} = (h_2,-h_1)$ and the reality condition $\gamma_{-\underline{k}} = -\bar{\gamma}_{\underline{k}}$ holds. By considering also the similar expansions for $\underline{f}$ and p,

$$\underline{f}(t) = \sum_{\underline{k}\neq 0} \left[f_{\underline{k}}(t) \frac{\underline{k}^{\perp}}{|\underline{k}|} + \tilde{f}_{\underline{k}}(t) \frac{\underline{k}}{|\underline{k}|} \right] e^{i\underline{k}\cdot\underline{x}}, \tag{2.3}$$

$$p(\underline{x},t) = \sum_{\underline{k}\neq 0} p_{\underline{k}}(t)\, e^{i\underline{k}\cdot\underline{x}}, \tag{2.4}$$

equations (2.1) yield the following system of ordinary differential equations:

$$\begin{cases} \dot{\gamma}_{\underline{k}} = -\nu|\underline{k}|^2\gamma_{\underline{k}} - i \displaystyle\sum_{\underline{k}_1+\underline{k}_2+\underline{k}=0} \frac{\left(\underline{k}_1^{\perp}\cdot\underline{k}_2\right)\left(k_2^2-k_1^2\right)}{2|\underline{k}_1||\underline{k}_2||\underline{k}|} \bar{\gamma}_{\underline{k}_1}\bar{\gamma}_{\underline{k}_2} + f_{\underline{k}} \\ \gamma_{\underline{k}}(0) = u_{ok}\,. \end{cases} \tag{2.5}$$

Once $\{\gamma_{\underline{k}}\}_{\underline{k}\neq 0}$ is obtained, $p_{\underline{k}}(t)$ is given by

$$p_{\underline{k}}(t) = \tilde{f}_{\underline{k}}(t) - i \sum_{\underline{k}_1+\underline{k}_2+\underline{k}=0} \frac{\left(\underline{k}_2\cdot\underline{k}_1^{\perp}\right)\left(\underline{k}_1\cdot\underline{k}_2^{\perp}\right)}{|\underline{k}_1||\underline{k}_2||\underline{k}|} \bar{\gamma}_{\underline{k}_1}\bar{\gamma}_{\underline{k}_2} \;. \tag{2.6}$$

Equations (2.5) describe the motion of infinitely many coupled gyroscopes, subject to friction forces $-\nu|\underline{k}|^2 \gamma_{\underline{k}}$ and driving forces $f_{\underline{k}}$ (see Gallavotti [10] and Boldrighini [11]).

Now, let us consider a set L consisting of N wave vectors $\underline{k}_1, \underline{k}_2, \ldots, \underline{k}_N$ placed in the half plane

$$\Pi^+ = \{(x,y)\ ,\ x \geq 0,\ y \geq 0 \quad \text{if} \quad x = 0\}\ ,$$

and their opposites $\underline{k}_{-j} = -\underline{k}_j$, $j = 1, \ldots, N$. The truncated Navier-Stokes equations are defined as

$$\begin{cases} \dot{\gamma}_{\underline{k}} = -\nu|\underline{k}|^2\gamma_{\underline{k}} - i \displaystyle\sum_{\substack{\underline{k}_1+\underline{k}_2+\underline{k}=0 \\ \underline{k}_1,\underline{k}_2 \quad L}} \frac{\left(\underline{k}_1^{\perp}\cdot\underline{k}_2\right)\left(k_2^2-k_1^2\right)}{2|\underline{k}_1||\underline{k}_2||\underline{k}|} \bar{\gamma}_{\underline{k}_1}\bar{\gamma}_{\underline{k}_2} + f_{\underline{k}} \\ \gamma_{-\underline{k}} = -\bar{\gamma}_{\underline{k}} \\ \gamma_{\underline{k}}(0) = u_{o\underline{k}}\ ,\ k\in L\ . \end{cases} \tag{2.7}$$

This is a system of N differential equations in the complex functions $\gamma_j(t)$, $j = 1, \ldots, N$ (from now on we adopt the simpler notation γ_j for $\gamma_{\underline{k}_j}$ and γ_{-j} for $\gamma_{-\underline{k}_j}$). Such a system has the following form

$$\dot{\gamma}_j = -\nu a_j\gamma_j + i \sum_{\substack{m,n=-N \\ |m|\neq j \\ |n|\neq j \\ |m|<|n|}}^{N} b_{jmn}\,\bar{\gamma}_m\,\bar{\gamma}_n + f_j\ ,\ j = 1,\ldots,N, \tag{2.8}$$

where $a_j > 0$ and $b_{jmn} = 0$ if $\underline{k}_j + \underline{k}_m + \underline{k}_n \neq 0$.

By expressing the $\gamma_j(t)$ in the polar form

$$\gamma_j(t) = \rho_j(t)\, e^{i\theta_j(t)}\ ,\ j = 1,\ldots,N,$$

with $\rho_j(t)$ and $\theta_j(t)$ varying in $(-\infty, +\infty)$, system (2.8) can be rewritten separating its real and imaginary parts. In doing so, we obtain also a formulation of the system not depending on the vectors $-\underline{k}_j$; hence, from

now on we shall refer only to the vectors $\underline{k}_j \in \Pi^+$. If $f_j = R_j\, e^{i\eta_j}$, we find

$$\dot{\rho}_j = -\nu a_j\rho_j + \sum_{\substack{m,n=1 \\ m\neq j \\ n\neq j \\ m<n}} c_{jmn}\rho_m\rho_n \sin\left[(-1)^{\beta_j}\theta_j + (-1)^{\beta_m}\theta_m + (-1)^{\beta_n}\theta_n\right] + R_j\cos(\eta_j - \theta_j)\ , \tag{2.9a}$$

$$\rho_j\dot{\theta}_j = \sum_{\substack{m,n \\ m\neq j \\ n\neq j \\ m<n}}^{N} (-1)^{\beta_j} c_{jmn}\rho_m\rho_n \cos\left[(-1)^{\beta_j}\theta_j + (-1)^{\beta_m}\theta_m + (-1)^{\beta_n}\theta_n\right] + R_j \sin(\eta_j - \theta_j) , \tag{2.9b}$$

with $j = 1, \ldots, N$ and $|c_{jmn}| = \max_{\substack{m'=\pm m \\ n'=\pm n}} |b_{jm'n'}|$. The exponents β_j, β_m, β_n are determined by the geometrical connection occurring among the vectors $\underline{k}_j$, $\underline{k}_m$, $\underline{k}_n$. More precisely, since the three vectors are related by an equality of the kind $\underline{k}_{i_1} + \underline{k}_{i_2} = \underline{k}_{i_3}$, we have $\beta_{i_1} = \beta_{i_2} = 0$ and $\beta_{i_3} = 1$. Furthermore, if $j < m < n$, the relation $c_{jmn} + c_{mjn} + c_{njm} = 0$ holds. This is very important, because it guarantees that the solution of system (2.7) exists and is bounded for all time (see again References [10,11]).

Finally, after assuming $\nu = 1$ and, for simplicity, a force $\underline{f}$ independent of time t, we define the Reynolds number R as

$$R = \sum_{j}^{N} |f_j| = \sum_{j}^{N} R_j .$$

3) SYMMETRIES AND INVARIANT HYPERPLANES

The solution of truncated Navier-Stokes equations can be invariant with respect to suitable rotations. To investigate such a symmetry property, let us look for a nonzero constant vector $\underline{\alpha} = (\alpha_1, \ldots, \alpha_N)$, $0 \leq \alpha_i < 2\pi$, such that if $\{\gamma_j(t) = \rho_j(t)\, e^{i\theta_j(t)}\}_{j=1,\ldots,N}$ is a solution, then $\{\gamma_j(t) = \rho_j(t)\, e^{i[\theta_j(t)+\alpha_j]}\}_{j=1,\ldots,N}$ is also a solution.

In order for such an $\underline{\alpha}$ to exist, the following relations (modulo 2π) must hold

$$\alpha_j = 0 \ , \text{ for each } j \text{ such that } f_j \neq 0 \ ; \tag{3.1a}$$

$$\alpha_{j_1} + \alpha_{j_2} = \alpha_{j_3} \ , \text{ for each term } (j_1,j_2,j_3) \text{ such that } \underline{k}_{j_1} + \underline{k}_{j_2} = \underline{k}_{j_3} \ . \tag{3.1b}$$

These equalities provide a linear system of M_1+M_2 equations in the N unknowns α_i. M_1 is the number of the modes excited by the force $\underline{f}$ and M_2 is the number of the distinct triplets of vectors belonging to L and placed in Π^+ which form a triangle. Such a system admits either only the null solution or a continuum of solutions. In order to understand when this latter case occurs, let us consider the problem more closely. For every vector set L, there is a minimal subset L´ generating L in the following sense: L is obtainable via subsequent enlargements of L´ performed by adding to L´ every element of L-L´ which is either the sum or the difference of two vectors of L´. If L´ consists of N´ vectors, system (3.1b) can be reduced to N-N´ linearly independent equations, each of them expressing an α_i belonging to L-L´ as a linear combination of the α_i's belonging to L´. Because N´ $\geq$ 2, we always have symmetries when $\underline{f}$ acts only on one mode. With this hypothesis, and only with this one, the infinite system (2.5), for which

$N' = 2$ being $L'=\{(1,0),(0,1)\}$, admits a (linear) continuum of angular symmetries. If $N' = 2$ and $\underline{f}$ acts on more than one mode, no symmetry of this simple kind is present.

The numerical investigation of system (2.7), actually performed after writing it in the Cartesian form rather than in the polar one, is quite difficult and computationally time-consuming when two or more f_j's are different from zero and vary independently of each other. For this reason all the models studied up to now posited an external force $\underline{f}$ acting only on one mode. Then, let us assume $R_1 = R \neq 0$, $R_j = 0$, $j = 2, \ldots, N$. As we have just seen, in this case system (2.7) exhibits an infinite group of symmetries, one-dimensional or higher-dimensional depending on how the vector set L is chosen.

Looking at system (2.9) it is easy to see that part (a) gives the evolution of the $\rho_j(t)$'s while part (b) provides that of the $\theta_j(t)$'s. Furthermore, it is immediately observed that subsystem (2.9b), which is a nonautonomous system owing to its dependence on time through the ρ_j's, admits stationary solutions obtainable by solving the following linear system of equations (modulo π):

$$\begin{cases} \theta_1 = \eta_1 \\ \theta_{j_1} + \theta_{j_2} = \theta_{j_3} + \frac{\pi}{2}\ , \text{ for each term } (j_1,j_2,j_3) \text{ such that} \\ \qquad\qquad \underline{k}_{j_1} + \underline{k}_{j_2} = \underline{k}_{j_3}\ . \end{cases} \tag{3.2}$$

The same arguments advanced concerning system (3.1) are valid for this system. Then, we have a continuum of stationary solutions for the $\theta_j(t)$'s, each of them representing an N-dimensional hyperplane immersed in the 2N-dimensional phase space. All these hyperplanes are related to each other by the symmetry group.

These stationary solutions play an important role in the phenomenology of system (2.7). In fact there is numerical evidence, at least for R values not too large (we shall be more precise later), that every randomly chosen initial point tends to one of these infinitely many symmetric invariant hyperplanes. After the solution has been captured by such an N-dimensional subspace, the trajectory searches for another attractor in this new phase space. This latter attractor can be either a fixed point or a periodic orbit or a torus or a strange attractor. Each Fourier-component $\gamma_j(t)$, $j \neq 1$, tends in its complex plane to one of the infinitely many lines passing through the origin. $\gamma_1(t)$ always tends to the line defined by $\theta_1 = \eta_1$.

In all the models obtained from (2.7) and investigated before the present work, a particular solution has been studied, in which each component $\gamma_j(t)$ is either real or purely imaginary. Because such a particular solution leads to a system of N rather than 2N differential equations, the result is a gain in simplicity and a reduction in computational time. To consider such a "real" solution instead of the whole complex one is equivalent to taking the initial conditions directly on one of the infinitely many symmetric hyperplanes representing stationary solutions for system (2.9b).

Lastly, we remark that our analysis of system (2.9) has been limited to consideration of some simple properties. The existence of less obvious symmetries and more complicated invariant hypersurfaces cannot be excluded.

4) THE COMPLEX 5-MODE MODEL

As an example, let us consider the 5-mode model studied in [1,2] and re-examine it as complex model. Such a reviewing will make more evident what has been previously said and will allow us to make some interesting observations about the attractors, their location in the phase space and the bifurcations they undergo.

The complex 5-mode model, written in the form of system (2.9) with $\eta = 0$, is the following

$$\begin{cases} \dot\rho_1 = -2\rho_1 + 4\rho_2\rho_3 \sin(\theta_1-\theta_2+\theta_3) - 4\rho_4\rho_5 \sin(\theta_1-\theta_4+\theta_5) \\ \dot\rho_2 = -9\rho_2 + 3\rho_1\rho_3 \sin(\theta_1-\theta_2+\theta_3) \\ \dot\rho_3 = -5\rho_3 - 7\rho_1\rho_2 \sin(\theta_1-\theta_2+\theta_3) + R\cos\theta_3 \\ \dot\rho_4 = -5\rho_4 + \rho_1\rho_5 \sin(\theta_1-\theta_4+\theta_5) \\ \dot\rho_5 = -\rho_5 + 3\rho_1\rho_4 \sin(\theta_1-\theta_4+\theta_5) \\ \rho_1\dot\theta_1 = 4\rho_2\rho_3 \cos(\theta_1-\theta_2+\theta_3) - 4\rho_4\rho_5 \cos(\theta_1-\theta_4+\theta_5) \, , \\ \rho_2\dot\theta_2 = -3\rho_1\rho_3 \cos(\theta_1-\theta_2+\theta_3) \\ \rho_3\dot\theta_3 = -7\rho_1\rho_2 \cos(\theta_1-\theta_2+\theta_3) - R\sin\theta_3 \\ \rho_4\dot\theta_4 = -\rho_1\rho_5 \cos(\theta_1-\theta_4+\theta_5) \\ \rho_5\dot\theta_5 = 3\rho_1\rho_4 \cos(\theta_1-\theta_4+\theta_5) \end{cases} \tag{4.1}$$

This system has the two-parameter group of symmetries

$$S_{\alpha\beta} : \ (\theta_1,\theta_2,\theta_3,\theta_4,\theta_5) \to (\theta_1+\alpha,\theta_2+\alpha,\theta_3,\theta_4+\alpha+\beta,\theta_5+\beta), \text{ where}$$

$\alpha,\beta \in [0,2\pi)$.

The 5-dimensional hyperplanes representing stationary solutions for $(\theta_1(t), \theta_2(t), \ldots, \theta_5(t))$ are given by

$$\Pi_{\delta\varepsilon} \equiv (\theta_1 = \delta,\ \theta_2 = \delta - \tfrac{\pi}{2}\ ,\ \theta_3 = 0,\ \theta_4 = \delta + \varepsilon - \tfrac{\pi}{2}\ ,\ \theta_5 = \varepsilon)\ ,$$

with $\delta,\varepsilon \in [0,\pi)$.

System (4.1) has the following fixed points:

a) for $0 < R \le R_1 = 5\sqrt{\frac{3}{2}}$, $P_o \equiv (0,\ 0,\ \frac{R}{5}\ ,\ 0,\ 0)$;

b) for $R_1 < R \le R_2 = \frac{80}{9}\sqrt{\frac{3}{2}}$, P_o and

$$P_\alpha \equiv \left(\sqrt{\sqrt{6}(R - R_1)/7}\ e^{i\alpha}\ ,\ \sqrt{(R - R_1)/7\sqrt{6}}\ e^{i(\alpha-\frac{\pi}{2})}\ ,\ \sqrt{3/2},\ 0,\ 0\right);$$

c) for $R > R_2$, P_o, P_α and

$$P_{\alpha\beta} \equiv \left(\sqrt{\frac{5}{3}}\ e^{i\alpha}\ ,\ \frac{3}{80}\sqrt{\frac{5}{3}}\ e^{i(\alpha-\frac{\pi}{2})}\ ,\ \frac{9}{80}\,R\ ,\ \frac{1}{3}\sqrt{\left(\frac{9}{80}\,R\right)^2 - \frac{3}{2}}\ e^{i(\alpha+\beta-\frac{\pi}{2})}\ ,\right.$$
$$\left.\sqrt{\frac{5}{3}}\sqrt{\left(\frac{9}{80}\,R\right)^2 - \frac{3}{2}}\ e^{i\beta}\right),$$

where $\alpha,\beta \in [0,2\pi)$.

The fixed point P_o is stable for $0 < R < R_1$. At R_1 it becomes unstable bifurcating into the fixed points P_α which constitute a closed curve surrounding P_o. Looking at the eigenvalues of the Liapunov matrix at the bifurcation point, we see that two identical real eigenvalues cross the imaginary axis from left to right. A real eigenvalue becoming positive suggests a bifurcation into fixed points; a pair of conjugate eigenvalues acquiring positive real part suggests a bifurcation into a closed curve. In our degenerate case the fixed point bifurcates into a closed curve of fixed points, which appears to be the simplest behavior one can expect.

A similar bifurcation takes place at R_2. Becoming unstable each P_α generates a curve of fixed point $P_{\alpha\beta}$ surrounding it. All the $P_{\alpha\beta}$'s are situated on a two-dimensional torus. If we consider the curve formed by the P_α's, we see that a bifurcation of a closed curve into a torus T^2 occurs.

At $R_3 \cong 22.854$ the fixed points $P_{\alpha\beta}$ also become unstable, each of them bifurcating via a Hopf bifurcation into a closed orbit. This is the same as saying that the two-torus of fixed points bifurcates into a three-torus completely covered with closed periodic orbits. Because each of these periodic orbits originates a sequence of period-doubling bifurcations, it is straightforward to deduce that the three-torus undergoes a sequence of doubling bifurcations at each of which one of its frequencies halves (the other two are zero).

Numerical evidence indicates that the invariant hyperplanes $\Pi_{\delta\varepsilon}$ are stable and globally attracting for every Reynolds number R. In view of the symmetry group, this means that the behavior of the "real" model studied in [1,2], which corresponds to the case of taking the initial conditins on Π_{oo}, tells us everything about the behavior of the complex model.

The above remarks concerning the spatial allocation of the closed orbits which arise from the points $P_{\alpha\beta}$, and the fact that the $\Pi_{\delta\varepsilon}$'s are always attracting, lead us to posit an analogous toroidal structure also for all the other attractors, periodic orbits or strange attractors, present in the complex model.

5) ATTRACTIVITY OF THE INVARIANT HYPERPLANES

As has just been seen, there is numerical evidence that in the 5-mode model the hyperplanes representing stationary solutions for the $\theta_i(t)$'s are stable and attracting for every value of the Reynolds number R. This situation does not occur for the other larger systems we have considered, i.e. the 7-mode model studied in [4], the 8- and 9-mode ones investigated in [7]. For each of these models there is a not well-definable critical parameter value R_c beyond which the invariant hyperplanes appear to be not attracting. In fact, by considering the behavior for $R > R_c$ of $\gamma_1(t)$, the only Fourier-component excited by the external force, it can be seen that the associated $\theta_1(t)$ cannot obtain the constant asymptotic value η_1 (see system (3.2)). For R slightly greater than R_c, $\theta_1(t)$ fluctuates around η_1 with very small oscillations. As R increases, these oscillations become gradually larger and larger. Several experiments with Runge-Kutta methods of different order and different integration steps, yielding nearly the same results, seem to exclude numerical noise as the cause of instability of the solution.

The critical values R_c, roughly approximated, for the 7-, 8- and 9-mode models we have considered, are 73.5, 17 and 15.5 respectively. These numerical values have different implications depending on the model. In the two larger models the invariant hyperplanes become unattracting when the corresponding real models already exhibit turbulence. However, since it seems plausible that the flow would be more chaotic in these cases, it could be said that there is no significant difference between the complex and the real phenomenologies. In contrast, some difference exists between the phenomenologies of complex and real 7-mode models. In fact, for R = 73.5, when the hyperplanes no longer attract, the real model still shows a periodic behavior persisting up to R ≅ 248. This probably means that in the complex model turbulence takes place at an earlier time.

6) CONCLUSION

If the external force acts only on one mode, complex truncated Navier-Stokes equations always exhibit infinitely many invariant hyperplanes symmetrically placed in consequence of a continuum group of symmetries. A numerical study of some models, real versions of which have been studied before this work, shows two different behaviors. In a 5-mode model the invariant hyperplanes turn out to be globally attracting for every value of the Reynolds number R. Every random initial point is first captured by one of these hyperplanes and then tends to some attractor on it. By contrast, in all the other models we consider, there exists a critical value R_c, beyond which the invariant hyperplanes no longer attract. So, the real model, corresponding to the case in which all the initial points are taken on one of the invariant hyperplanes, displays the whole complex phenomenology only in the 5-mode system. In the other cases, when the parameter is larger than R_c, the fluctuating behavior of the $\theta_i(t)$'s suggests a more complicated phenomenology in the case of the complex model.

Interesting bifurcations can occur in the complex phase space. A stable fixed point, becoming unstable, can bifurcate into a closed curve consisting entirely of fixed points. In turn such a curve can bifurcate into a two-dimensional torus covered entirely by either fixed points or closed orbits. Two- or three-dimensional tori, covered all over by closed orbits, can undergo an infinite sequence of doubling bifurcations when the frequency of the closed orbits halves an infinite number of times.

ACKNOWLEDGMENTS

I gratefully acknowledge many conversations with G. Gallovotti, whose suggestions have been very useful. I am also grateful to J. D. Farmer and M. J. Feigenbaum for their interest in this work and to E. Jen for improving my English considerably. It is a pleasure to thank the Center for Nonlinear Studies, Los Alamos National Laboratory, for its kind hospitality and for providing computer facilities.

REFERENCES

[1] C. Boldrighini - V. Franceschini, "A Five-Dimensional Truncation of the Plane Incompressible Navier-Stokes Equations," Commun. Math. Phys., 64 (1979) 159.

[2] V. Franceschini - C. Tebaldi, "Sequences of Infinite Bifurcations and Turbulence in a Five-Mode Truncation of the Navier-Stokes Equations," J. Stat. Phys., 21 (1979) 707.

[3] P. M. Angelo - G. Riela, "A Six-Mode Truncation of the Navier-Stokes Equations on a Two-Dimensional Torus: a Numerical Study," Il Nuovo Cimento, 64B (1981) 207.

[4] V. Franceschini - C. Tebaldi, "A Seven-Mode Truncation of the Plane Incompressible Navier-Stokes Equations," J. Stat. Phys., 25 (1981) 397.

[5] L. Tedeschini-Lalli, "Truncated Navier-Stokes Equations: Continuous Transition from a Five-Mode to a Seven-Mode Model," J. Stat. Phys., to appear.

[6] V. Franceschini, "Typical Behaviours in Dissipative Differential Equations," Rend. Sem. Mat. Univ. e Polit. Torino, to appear.

[7] V. Franceschini, "Two Models of Truncated Navier-Stokes Equations on a Two-Dimensional Torus," preprint.

[8] M. J. Feigenbaum, "Quantitative Universality for a Class of Nonlinear Transformations," J. Stat. Phys., 19 (1978) 25.

[9] M. J. Feigenbaum, "The Transition to Aperiodic Behavior in Turbulent Systems," Commun. Math. Phys., 77 (1980) 65.

[10] G. Gallavotti, "Meccanica Elementare," Boringhieri (1980).

[11] C. Boldrighini, "Introduzione alla Fluidodinamica," Quaderni dei gruppi di ricerca del C.N.R., 1979.

COUPLED NONLINEAR OSCILLATORS
J. Chandra and A.C. Scott (eds.)

FREQUENCY LOCKING IN JOSEPHSON POINT CONTACTS

Patrick S. Hagan

Exxon Corporate Research Science Laboratory
P.O. Box 45
Linden, NJ 07036

We consider a dc-current-driven Josephson point contact shunted by a resistive element. We calculate the voltage across the contact when a small-amplitude ac signal current with frequency ω is applied. We find that resonances occur at first order when ω is near the Josephson frequency ω_o, and that resonances occur at second order when ω is near $\omega_o/2$. By using singular perturbation techniques, we find accurate expressions for the contact voltage, even when one of the above resonances occurs. These expressions exhibit frequency (phase) locking when ω is near enough to either ω_o or $\omega_o/2$, and frequency pulling otherwise. The regions where frequency locking occurs are obtained.

1. INTRODUCTION

The Josephson junction has many practical applications for detecting weak high-frequency electromagnetic signals, so quantitative analyses of the junction's response to applied signals are very useful. Previously, numerical [1] and analog [2] computational methods have been used to provide quantitative explanations for experimental observations. Also, analytic techniques - in particular regular perturbation methods [3] - have been used to obtain quantitative results. However, regular perturbation methods fail when resonance (and the corresponding effects of frequency locking and frequency pulling) occurs. In this article we use a singular perturbation technique to obtain expressions for the response of a point junction which are accurate even when resonance occurs [4].

Specifically, we will find the voltage response of a dc-current-driven Josephson point contact shunted by a resistive element to a weak applied ac signal current. First, we will derive the equation governing the Josephson point contact in Section 2. Then, this circuit equation will be analyzed in Section 3. There we will find that the ac signal and the Josephson junction resonate at first order when the signal frequency ω is near $\pm\omega_o$, where ω_o is the Josephson frequency. A singular perturbation technique is used to obtain an expression which accurately approximates the contact voltage, even when ω is near $\pm\omega_o$. In Section 4 we will examine this expression for the contact voltage. We will find that this expression shows that frequency locking (alternatively called phase locking or entrainment) of the Josephson junction to the ac signal occurs when ω is in a small interval about ω_o. When ω is near to (but not in) this interval, the oscillation frequency of the junction is "pulled" toward the signal frequency. Finally, when ω is not near ω_o our expression for the contact voltage reduces to the expression obtained by using a regular perturbation expansion [3]. The resonance phenomena of frequency locking and frequency pulling have been observed experimentally [3].

We will briefly discuss the effects of second and higher order resonances in Section 5. For example, when ω is near enough to $\pm\omega_o/2$ the Josephson junction frequency locks to the second harmonic of the ac signal due to a second order resonance. Some experimental observations of this interaction of the second harmonic with the junction have been reported [3].

2. THE GOVERNING EQUATION

We now briefly derive the circuit equation which governs the Josephson point contact. The equivalent circuit of a current driven point contact carrying both pair and quasiparticle current is shown in Fig. 1. Following the derivation in

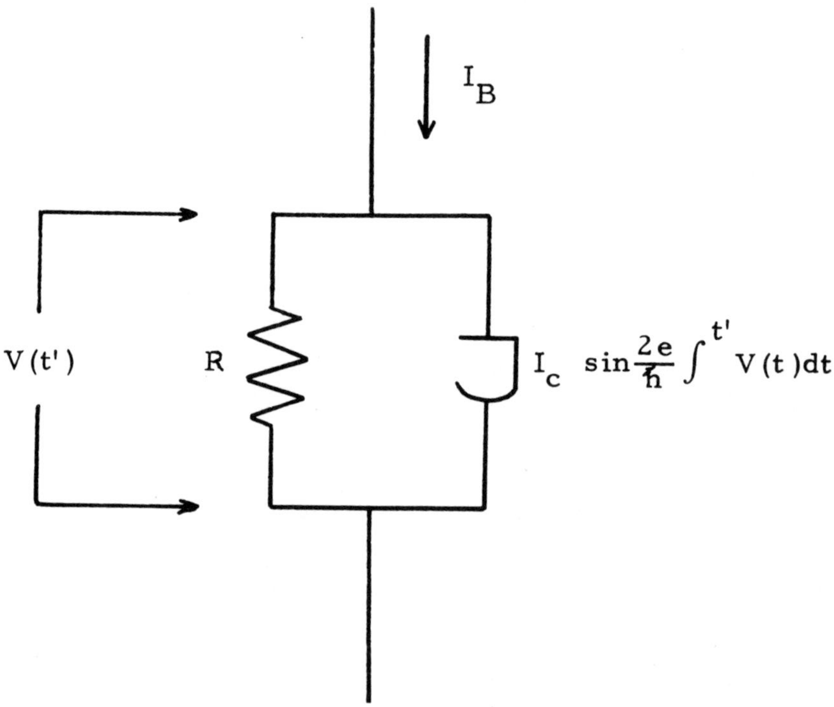

Figure 1
Equivalent circuit of current-driven Josephson point contact.

[3] we neglect all spatial variations of the pair current density and we neglect all reactive elements in the circuit. This yields a pair current

$$I_p = I_c \sin \phi(t') , \tag{2.1}$$

where

$$\phi(t') = \frac{2e}{\hbar} \int^{t'} V(s)ds , \tag{2.2}$$

and where I_c is the critical super-current, V is the voltage across the junction, and t' is time. The circuit equation is

$$V/R + I_c \sin\phi = I_B , \tag{2.3}$$

where I_B is the battery current and R is the normal leakage resistance.

We now define the dimensionless variables

$$A = \frac{I_B}{I_c} , \; v = \frac{V}{RI_c} , \; t = \frac{2e}{\hbar} RI_c t' . \tag{2.4}$$

In terms of these variables, the circuit Eq. (2.3) is

$$\frac{d\phi}{dt} + \sin\phi = A . \tag{2.5}$$

When a small ac signal current $I_s \sin\omega t$ is added, then the circuit equation clearly becomes

$$\frac{d\phi}{dt} + \sin\phi = A + \varepsilon \sin\omega t , \tag{2.6}$$

where $\varepsilon \equiv I_s/I_c$ and where we assume that $0 < \varepsilon << 1$. Equation (2.6) is the equation which governs our point junction. We will analyze the behavior of the solutions of (2.6) as ε and ω vary, in particular focusing our attention on the resonance solutions.

3. FIRST-ORDER RESONANCES

We now analyze (2.6) to first order in ε. When $\varepsilon = 0$, Eq. (2.6) reduces to

$$\frac{d\phi}{dt} + \sin\phi = A . \tag{3.1}$$

We assume that $A > 1$; i.e., that the battery current is supra-critical. Then it is easy to show that the general solution of (3.1) is

$$\phi(t) = \Phi(t-t_o) , \tag{3.2}$$

where t_o is an arbitrary constant and where the function $\Phi(\theta)$ is defined by

$$\tan\frac{1}{2}\,[\Phi(\theta)-\pi/2] \equiv \sqrt{\frac{A-1}{A+1}}\,\tan\frac{1}{2}\,\omega_o\theta . \tag{3.3}$$

Here $\omega_o \equiv \sqrt{A^2-1}$ is the Josephson frequency.

We now assume that the ac signal amplitude is small; i.e., that $\varepsilon << 1$. Now typically one would solve 2.6 by substituting the expansion

$$\phi(t) = \phi^{(0)}(t) + \varepsilon\phi^{(1)}(t) + \varepsilon^2\phi^{(2)}(t) + \ldots , \tag{3.4}$$

into 2.6, and then solving for the coefficients $\phi^{(0)}$, $\phi^{(1)}$, $\phi^{(2)}$,... . However, this regular perturbation method fails when resonance occurs, because then some terms in (3.4) are infinite. We will avoid this problem by using a multi-scale expansion.

As a first step, we will greatly simplify all subsequent algebra by transforming the circuit Eq. (2.6) to an equation with simpler structure. Specifically, we define a new <u>dependent</u> variable $\theta(t)$ by the implicit relationship

$$\phi(t) = \Phi(\theta(t)) , \tag{3.5}$$

where Φ is the function defined in Eq. (3.3). We now switch from using ϕ as our dependent variable to using θ. We substitute (3.5) and (3.3) into (2.6), and after some algebra we obtain

$$\frac{d\theta}{dt} = 1 + \frac{\varepsilon}{\omega_o^2} \{A \sin \omega t+\frac{1}{2} \sin(\omega t+\omega_o\theta)+\frac{1}{2} \sin(\omega t-\omega_o\theta)\} . \tag{3.6}$$

Equation (3.6) is exact; no approximations have been used in reducing (2.6) to (3.6). Although Eq. (3.6) looks more complicated than (2.6), the linearity of (3.6) at $\varepsilon = 0$ makes this equation very easy to work with.

We now use a multi-scale perturbation method to solve (3.6). We define the "slow time" variable T by $T = \varepsilon t$, we expand $\theta(t,\varepsilon)$ as

$$\theta(t,\varepsilon) = \theta^{(0)}(t,T) + \varepsilon\theta^{(1)}(t,T) + \varepsilon^2\theta^{(2)}(t,T) + \dots , \tag{3.7}$$

and we require that $\theta^{(1)}, \theta^{(2)},\dots$ be of a size smaller than $0(1/\varepsilon)$ for all t. Next, we substitute $d/dt = \partial/\partial t + \varepsilon\partial/\partial T$ and the expansion (3.7) into (3.6), and we equate the coefficients of like powers of ε. This yields the following sequence of equations

$$\frac{\partial\theta^{(0)}}{\partial t} = 1 , \tag{3.8a}$$

$$\frac{\partial\theta^{(1)}}{\partial t} = - \frac{\partial\theta^{(0)}}{\partial T} + \frac{1}{\omega_o^2} \{A \sin \omega t+\frac{1}{2} \sin(\omega t+\omega_o\theta^{(0)})+\frac{1}{2} \sin(\omega t-\omega_o\theta^{(0)})\} , \tag{3.8b}$$

$$\vdots$$

The general solution of (3.8a) is clearly

$$\theta^{(0)} = t + \beta(T) , \tag{3.9a}$$

where the function $\beta(T)$ is undetermined at this stage of the perturbation scheme. We substitute (3.9a) into (3.8b) and solve for $\theta^{(1)}$, obtaining

$$\theta^{(1)} = - t \frac{\partial\beta}{\partial T} - \frac{1}{\omega_o^2} \{\frac{A}{\omega} \cos \omega t + \frac{\cos[(\omega+\omega_o)t+\omega_o\beta]}{2(\omega+\omega_o)} + \frac{\cos[(\omega-\omega_o)t-\omega_o\beta]}{2(\omega-\omega_o)}\} + K , \tag{3.9b}$$

where K is an integration constant. Now if ω is <u>not</u> near ω_o, $-\omega_o$, or 0, (i.e., if resonance does <u>not</u> occur at first order), then the requirement that $\theta^{(1)}$ must never become unreasonably large implies that $d\beta/dT = 0$. Thus, for this case $\beta(T)$ is some constant β_o. In summary, an expression for ϕ accurate to order ε and for times t no larger than $0(1/\varepsilon)$ is

$$\phi(t,\varepsilon,\omega) \equiv \Phi(\theta(t,\varepsilon,\omega)) , \tag{3.10a}$$

where $\Phi(\theta)$ is defined by (3.3) and where $\theta(t,\varepsilon,\omega)$ is given by

$$\theta(t,\varepsilon,\omega) = t + \beta_o - \frac{\varepsilon}{\omega_o^2} \{\frac{A}{\omega} \cos \omega t + \frac{\cos[(\omega+\omega_o)t+\omega_o\beta_o]}{2(\omega+\omega_o)} + \frac{\cos[(\omega-\omega_o)t-\omega_o\beta_o]}{2(\omega-\omega_o)}\}$$

$$+ \varepsilon K + 0(\varepsilon^2) , \omega \neq \pm \omega_o , \omega \neq 0 . \tag{3.10b}$$

This expression agrees with the result in [3] to order ε. We will briefly examine the solution for the non-resonance case in Section 4.

We now treat the cases where resonance occurs. First, suppose that $\omega \approx \omega_o$. We define η by

$$\omega \equiv \omega_o + \eta\varepsilon \ , \tag{3.11}$$

we substitute this relation and (3.9a) into (3.8b), and we solve for $\theta^{(1)}$. In this manner we obtain

$$\theta^{(1)} = t\{\frac{1}{2\omega_o^2}\sin(\eta T - \omega_o\beta) - \frac{d\beta}{dT}\} - \frac{1}{\omega_o^3}\{A\cos(\omega_o t + \eta T) + \frac{1}{4}\cos(2\omega_o t + \eta T + \omega_o\beta(T))\}$$

instead of (3.9b), where K is again an integration constant. Since we require $\theta^{(1)}$ to remain reasonably sized for all t, we conclude that

$$\frac{d\beta}{dT} = \frac{1}{2\omega_o^2}\sin(\eta T - \omega_o\beta) \ . \tag{3.13}$$

Thus, for this resonance case we find that to leading order in ε the solution is $\theta = t + \beta(T)$, where β evolves on the slow time scale $T = \varepsilon t$ according to (3.13). Equation (3.13) is the key equation which describes the $O(1)$ effects of the resonant $O(\varepsilon)$ signal.

The general solution of (3.13) is

$$\beta(T) = \{\eta T - P(T)\}/\omega_o \ , \tag{3.14a}$$

where P(T) is given by

$$\gamma \tan\frac{1}{2}P(T) = \begin{cases} \sqrt{\gamma^2-1}\tan\{\sqrt{\gamma^2-1}\,\frac{T-T_o}{4\omega_o}\} + 1 & \text{if} \quad \gamma^2 \geq 1 \\ -\sqrt{1-\gamma^2}\tanh\{\sqrt{1-\gamma^2}\,\frac{T-T_o}{4\omega_o}\} + 1 & \text{if} \quad \gamma^2 \leq 1 \\ -\sqrt{1-\gamma^2}\,\mathrm{ctnh}\{\sqrt{1-\gamma^2}\,\frac{T-T_o}{4\omega_o}\} + 1 & \text{if} \quad \gamma^2 \leq 1 \end{cases} \ , \tag{3.14b}$$

and where the definition

$$\gamma \equiv 2\eta\omega_o \equiv 2\omega_o(\omega - \omega_o)/\varepsilon \tag{3.14c}$$

has been used. In particular, the initial value of β determines not only the integration constant T_o, but also determines whether the tanh or the ctnh solution is to be used when $\gamma^2 \leq 1$. Specifically, when $\gamma^2 \leq 1$, the tanh solution is used when $|1 + \gamma \tan 1/2\, \omega_o\bar{\beta}(o)| < \sqrt{1-\gamma^2}$; otherwise the ctnh solution is used.

In summary, for the case of $\omega \quad \omega_o$, an expression for ϕ accurate to order ε and for times t no larger than $O(1/\varepsilon)$ is

$$\phi(t,\varepsilon,\omega) = \Phi(\theta(t,\varepsilon,\omega)) \ , \tag{3.15a}$$

where $\Phi(\theta)$ is given by (3.3), where $\theta(t,\varepsilon,\omega)$ is given by

$$\theta(t,\varepsilon,\omega) = t + \beta(T) - \frac{\varepsilon}{\omega_o^3}\{A\cos(\omega_o t + \eta T) + \frac{1}{4}\cos(2\omega_o t + \eta T + \omega_o\beta(T))\}$$

$$+ \varepsilon K + O(\varepsilon^2) \ , \tag{3.15b}$$

and where $\beta(T)$ is given by (3.14a-3.14c). We will examine this resonance solution in Section 4. There we will find that as γ^2 becomes large (as ω shifts away from ω_o) this expression reduces to the expression (3.10b) for the nonresonance case, as γ^2 approaches unity frequency "pulling" and rhythm splitting occur, and finally when $\gamma^2 < 1$ frequency locking occurs.

First order resonance also occurs if $\omega \approx -\omega_o$. This case is virtually identical to the $\omega \approx +\omega_o$ case. Indeed, the solution for $\phi(t,\varepsilon,\omega)$ when $\omega \approx -\omega_o$ can be obtained by replacing ω_o by $-\omega_o$ in Eqs. (3.14a-3.15b).

The other case of first-order resonance is $\omega \approx 0$. For this case we define μ by $\omega \equiv \mu\varepsilon$, we substitute this expression and (3.9a) into (3.8b), and we solve for $\theta^{(1)}$. In this way we find that when $\omega \approx 0$,

$$\theta(t,\varepsilon,\omega) = t + \beta(T) + \frac{\varepsilon}{\omega_o^3} \sin \mu T \cdot \sin(\omega_o t+\omega_o\beta(T)) + \varepsilon K + O(\varepsilon^2) , \tag{3.16a}$$

where

$$\beta(T) = \frac{A}{\mu\omega_o^2} \{1-\cos \mu(T-T_o)\} . \tag{3.16b}$$

This case can be viewed as the battery current being $A + \varepsilon \sin \mu T$, and thus fluctuating on an adiabatically long time-scale. A careful examination shows that (3.16a), (3.16b) represent the first-order effects of the slow variation of the Josephson frequency $\omega_o(T) \equiv \{(A+\varepsilon \sin \mu T)^2 - 1\}^{\frac{1}{2}}$.

4. NATURE OF THE RESONANCE SOLUTION

We now examine the qualitative behavior of the non-resonance solution (3.10a), (3.10b) and the resonance solution (3.14a)-(3.15b). Although the variable θ is mathematically convenient, it cannot be easily interpreted physically. Thus we will discuss the solutions of the preceding section in terms of $v \equiv d\phi/dt$, which is the contact voltage in reduced units. Now

$$v \equiv \frac{d\phi}{dt} = \left[\Phi'(\theta^{(0)})+\varepsilon\theta^{(1)}\Phi''(\theta^{(0)})+\dots\right] \cdot \left[\frac{\partial\theta^{(0)}}{\partial t} + \varepsilon \frac{\partial\theta^{(1)}}{\partial t} + \frac{\partial\theta^{(0)}}{\partial T}) + \dots\right] . \tag{4.1}$$

Therefore

$$v = v^{(0)} + \varepsilon v^{(1)} + \varepsilon^2 v^{(2)} + \dots , \tag{4.2a}$$

where

$$v^{(0)} = \frac{\omega_o^2}{A + \cos[\omega_o t+\omega_o\beta(T)]} \tag{4.2b}$$

$$v^{(1)} = \frac{\omega_o^2}{A + \cos[\omega_o t+\omega_o\beta(T)]} \left(\frac{d\beta}{dT} + \frac{\partial\theta^{(1)}}{\partial t}\right) + \theta^{(1)} \frac{\omega_o^3 \sin[\omega_o t+\omega_o\beta(T)]}{(A + \cos[\omega_o t+\omega_o\beta(T)])^2} , \tag{4.2c}$$

and where $\theta^{(0)} + \varepsilon\theta^{(1)}$ is given by (3.10b) for the non-resonance case and by (3.15b) for the resonance case.

To interpret these solutions, we will look at the two key quantities of the first-order voltage response $\varepsilon v^{(1)}$ and the phase $\chi \equiv \omega_o t + \omega_o \beta(T)$ of the dominant term $v^{(0)}$ of the voltage expansion. For the non-resonance solution, the first order voltage response is

$$\varepsilon v^{(1)}(t) = \frac{\partial}{\partial t} F \ , \tag{4.3a}$$

where

$$F = -\varepsilon \left(\frac{A}{\omega} \cos \omega t + \frac{\cos[(\omega+\omega_o)t+\omega_o\beta_o]}{2(\omega+\omega_o)} + \frac{\cos[(\omega-\omega_o)t-\omega_o\beta_o]}{2(\omega-\omega_o)} - K \right) \times$$
$$[A + \cos \omega_o(t+\beta_o)]^{-1} \ . \tag{4.3b}$$

Although this response voltage varies in a complicated manner, averaging it over t yields

$$\overline{\varepsilon v^{(1)}} \equiv \lim_{t\to\infty} \frac{1}{t} \int_o^t \varepsilon v^{(1)}(s)ds = 0 \ . \tag{4.4}$$

Thus the time-averaged first-order voltage response is zero for the non-resonance case. Additionally, since $\beta(T)$ is a constant β_o for this case, the phase χ is

$$\chi = \omega_o t + \omega_o \beta_o \ , \tag{4.5}$$

so clearly the dominant voltage $v^{(0)}$ oscillates at precisely the Josephson frequency ω_o.

For the resonance case $\omega \approx \omega_o$, we substitute expressions (3.14a)-(3.14c) and (3.15b) into (4.2c), and we again find that $\varepsilon v^{(1)}(t,T)$ varies in a complicated manner. However, if we average $\varepsilon v^{(1)}(t,T)$ over t and T, we find that

$$\overline{\varepsilon v^{(1)}} \equiv \lim_{T\to\infty} \frac{1}{T} \int_o^T \left[\lim_{t\to\infty} \frac{1}{t} \int_o^t \varepsilon v^{(1)}(s,S)ds \right] dS \tag{4.6}$$

$$= \begin{cases} \varepsilon \,\mathrm{sgn}(\gamma) \dfrac{|\gamma| - \sqrt{\gamma^2-1}}{2\omega_o} & \text{if } \gamma^2 \geq 1 \\[2ex] \dfrac{1}{2} \varepsilon\gamma/\omega_o & \text{if } \gamma^2 \leq 1 \end{cases} \ , \tag{4.7}$$

where $\gamma \equiv 2\omega_o \eta = 2\omega_o(\omega-\omega_o)/\varepsilon$.

From (3.14a) we find that the phase of the dominant voltage term is

$$\chi = \omega_o t + \eta T - P(T) = \omega t - P(T) \ , \tag{4.8}$$

where P(T) is given by (3.14b), (3.14c). We note that tanh and ctnh both have the limiting value 1 as $T \to \infty$, so P(T) approaches a limiting value as $T \to \infty$, and thus χ goes to $\chi = \omega t$ + constant as T gets large. Thus we conclude that when $\gamma^2 \leq 1$ (when $|\omega-\omega_o| < \varepsilon/2\omega_o$), the Josephson junction is frequency locked to (entrained by) the ac signal. For the case of $\gamma^2 > 1$, we define the instantaneous frequency by $d\chi/dt = \partial\chi/\partial t + \varepsilon\, \partial\chi/\partial T$. Then when $\gamma^2 > 1$ the instantaneous frequency is

$$\frac{d\chi}{dt} = \omega_o + \frac{\varepsilon}{2\omega_o}\left\{\gamma - (\gamma^2-1)\Big/\left(\gamma + \cos\left[\sqrt{\gamma^2-1}\,\frac{T-T_o}{2\omega_o} - \alpha\right]\right)\right\}, \tag{4.9a}$$

where α is defined by

$$\cos\alpha = 1/\gamma\ ,\ \sin\alpha = \sqrt{\gamma^2-1}/\gamma\ . \tag{4.9b}$$

Thus, the instantaneous frequency oscillates on the εt time-scale with period $4\pi\delta_o/\sqrt{\gamma^2-1}$. Since the time-averaged instantaneous frequency is

$$\overline{\frac{d\chi}{dt}} = \omega_o + \frac{\varepsilon}{2\omega_o}\,\mathrm{sgn}(\gamma)\{|\gamma| - \sqrt{\gamma^2-1}\}\ , \tag{4.10}$$

which has the limits $\omega_o + \gamma\varepsilon/2\omega_o = \omega$ as $\gamma \to \pm 1$ and the limit ω_o as $\gamma^2 \to \infty$, the interpretation of frequency pulling of junction by the ac signal seems natural when $\gamma^2 > 1$.

In summary, for the non-resonance case the averaged first-order voltage response, $\varepsilon v^{(1)}$, is zero and the instantaneous frequency of the dominant voltage oscillation, $d\chi/dt$, is exactly the Josephson frequency ω_o. We have graphed $v^{(0)}(t)$ and $\varepsilon v^{(1)}(t)$ for the non-resonance case in Figs. 2 and 3. As ω approaches ω_o resonance phenomena begin to appear. Namely, as ω nears ω_o, $\varepsilon v^{(1)}$ grows according to

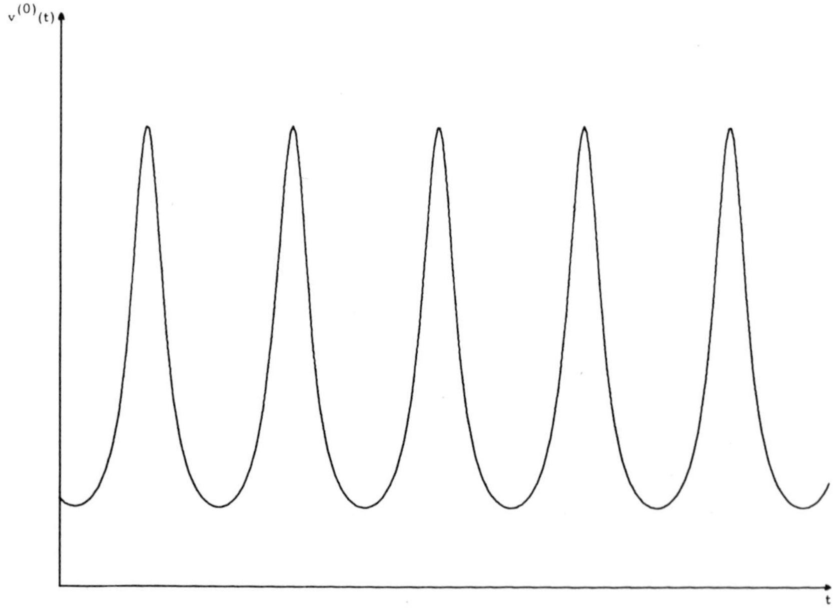

Figure 2
The primary voltage oscillation $v^{(0)}(t)$ of the junction when resonance does not occur.

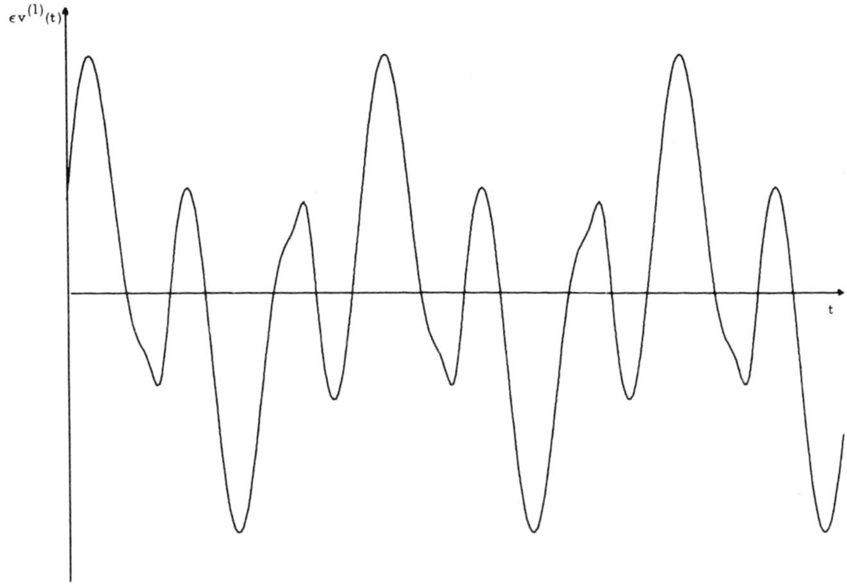

Figure 3

The first-order response voltage $\varepsilon v^{(1)}(t)$ when resonance does not occur.

formula (4.7) and $d\chi/dt$ acquires a small slow oscillation about its mean value given by (4.10). As ω nears ω_o, the period of the oscillation of $d\chi/dt$ increases, and it becomes infinite when $|\omega-\omega_o| = \varepsilon/2\omega_o$. When ω is near enough to ω_o (when $|\omega-\omega_o| < \varepsilon/2\omega_o$), then $\varepsilon v^{(1)}$ is given by $\omega-\omega_o$ and the quantity $d\chi/dt$ locks onto the value ω. In Fig. 4 we have graphed $d\chi/dt$ when ω is near to (but outside of) the locking region $|\omega-\omega_o| < \varepsilon/2\omega_o$. In Fig. 5 we have illustrated $d\chi/dt$ when ω is inside the locking region. Finally $\varepsilon v^{(1)}$ is illustrated in Fig. 6. In Figs. 2-6 we have used the illustrative (but atypical) parameter values $A = \sqrt{2}$, $\omega_o = 1$. Also, we have used $\omega = 1.5$ for Figs. 2 and 3, $\omega = 1.11$ for Fig. 4, and $\omega = 1.08$ for Fig. 5.

5. SECOND AND HIGHER ORDER RESONANCES

The two-timing procedure can be extended to as high an order in ε as one wishes, although the algebra is soon overwhelming. One finds that the frequency of the primary voltage oscillation, ω^*, is an even power series in ε

$$\omega^* = \omega_o + \frac{\varepsilon^2}{4\omega_o(\omega^2-\omega_o^2)} + \varepsilon^4 \omega_4 + \dots \tag{5.1}$$

when no resonance occurs. Moreover resonance occurs at second order when ω $\omega^*/2$, and at third order when ω $\omega^*/3$. When $|\omega^*-2\omega| \leq A\varepsilon^2/2\omega_o^3$, the frequency of the junction's voltage oscillation is exactly 2ω; i.e., the junction frequency locks onto the ac signal's second harmonic. Similarly, the junction locks onto

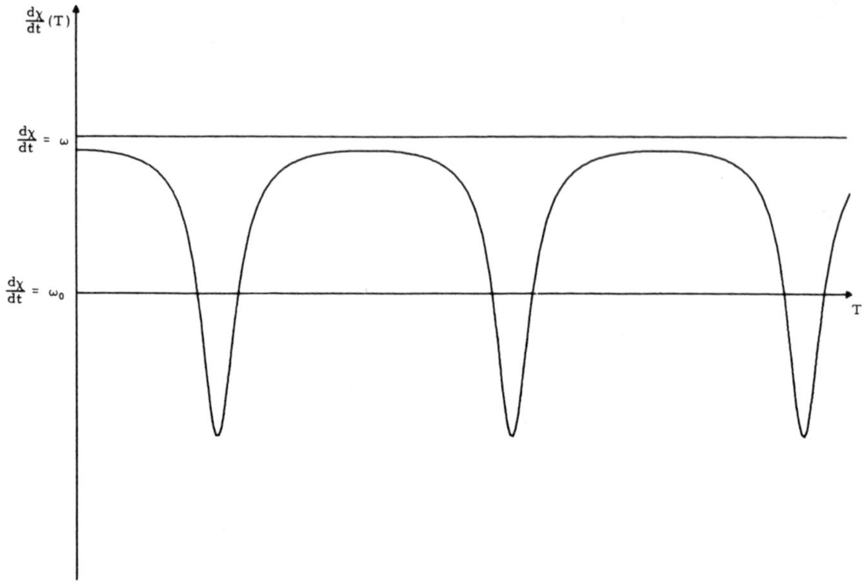

Figure 4
The instantaneous frequency $d\chi/dt$ as a function of the slow time variable $T = \varepsilon t$. Here the ac signal frequency ω is very near to (but outside of) the locking region $|\omega-\omega_o| < \varepsilon/2\omega_o$.

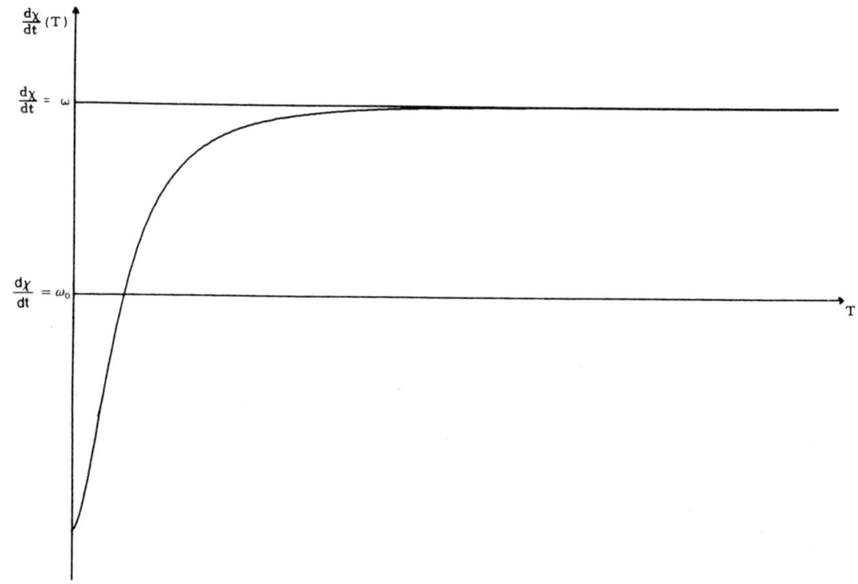

Figure 5
The instantaneous frequency $d\chi/dt$ as a function of T. Here the ac signal frequency ω is within the locking region $|\omega-\omega_o| < \varepsilon/2\omega_o$.

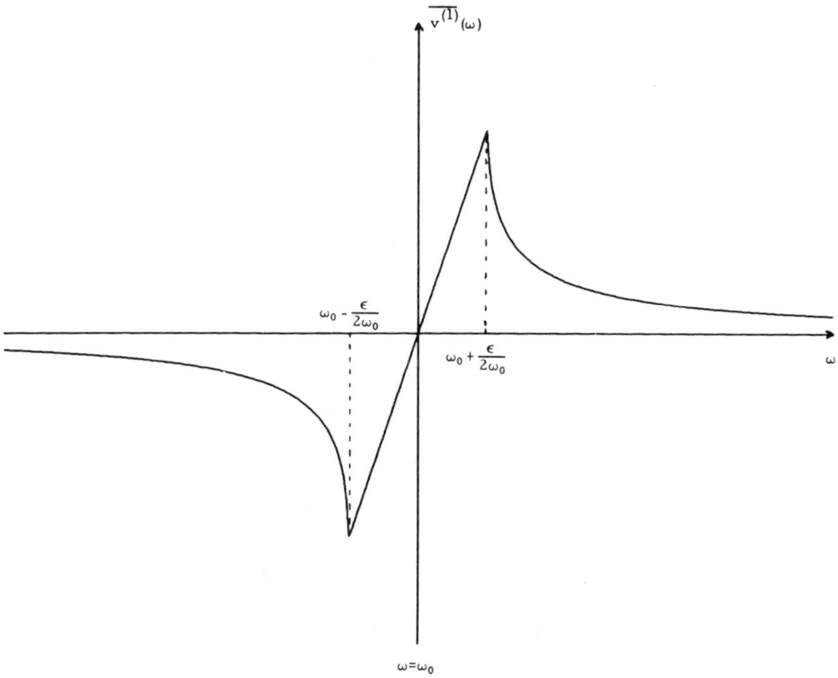

Figure 6

The time-averaged first-order response $\overline{\varepsilon v^{(1)}}$ versus signal frequency ω.

the signal's third harmonic when $|\omega^{*}-3\omega| < 9/128\ \varepsilon^3\ 1+8A^2/\omega_o^5$. When ω is near to (but outside of) these frequency locking regions, the voltage oscillation will be frequency "pulled" toward the appropriate signal harmonic.

REFERENCES

1. D. B. Sullivan, R. L. Peterson, V. Kose, and J. E. Zimmerman, J. Appl. Phys. 41, 4865 (1970).

2. N. R. Werthamer and S. Shapiro, Phys. Rev. 164, 523 (1967).

3. H. Kanter and F. L. Vernon, Jr., J. Appl. Phys. 43, 3174 (1972).

4. P. S. Hagan and D. S. Cohen, J. Appl. Phys. 50, 5408 (1979).

COUPLED NONLINEAR OSCILLATORS
J. Chandra and A.C. Scott (eds.)

SOLITON EXCITATIONS IN JOSEPHSON TUNNEL JUNCTIONS

P. S. Lomdahl*, O. H. Soerensen§, and P. L. Christiansen

Laboratory of Applied Mathematical Physics
The Technical University of Denmark
DK-2800 Lyngby, Denmark

A detailed numerical study of a sine-Gordon model of the Josephson tunnel junction is compared with experimental measurements on junctions with different L/λ_J ratio. The soliton picture is found to apply well on both relative long ($L/\lambda_J = 6$) and intermediate ($L/\lambda_J = 2$) junctions. We find good agreement for the current-voltage characteristics, power output, and for the shape and height of the zero field steps (ZFS). Two distinct modes of soliton oscillations are observed. (i) a bunched or congealed mode giving rise to the fundamental frequency f_1 on all ZFS's and (ii) a 'symmetric' mode which on the Nth ZFS yields the frequency Nf_1. Coexistence of two adjacent frequencies is found on the 3rd ZFS of the longer junction ($L/\lambda_J = 6$) in a narrow range of bias current as also found in the experiments. Small asymmetries in the experimental environment, a weak magnetic field, e.g., is introduced via the boundary conditions of our numerical model. This gives a junction response to variations in the applied bias current close to that observed experimentally.

I. INTRODUCTION

The study of soliton dynamics in connection with large Josephson tunnel junctions has recently drawn considerable theoretical[1-6] and experimental[7,8] attention. Fulton and Dynes[1] conceived the idea that the Josephson tunnel junction could support the resonant propagation of a soliton (or fluxon) trapped in the junction. The soliton being a 2π-jump in the phase difference (ϕ) across the insulating barrier which separates the two superconductors. The moving soliton is accompanied by a voltage pulse ($\sim \phi_t$) which can be detected at either end of the junction. The dc-manifestation of the motion is a sequence of equidistantly spaced branches in the current-voltage characteristic of the junction. These nearconstant voltage branches which were first reported by Chen, Finnegan and Langenberg[9] are known as zero-field steps (ZFS) because they occur in the absence of an external magnetic field. In contrast, the so-called Fiske steps are found only when a magnetic field is applied.[10] The ZFS's appear at voltages given by $\bar{V} = N\Phi_o \bar{c}/L$ where N is an integer, the flux quantum $\Phi_o = h/2e = 2.064 \cdot 10^{-15}$ Wb, $\bar{c}$ is the electromagnetic wave velocity in the junction and

*Address after 1 March 1982:
Center for Nonlinear Studies, Los Alamos National Laboratory, University of California, Los Alamos, NM 87545, U. S. A.

§Present address: Danish Post & Telegraph's Telecommunication's Laboratory, Agade 154, 2200 Copenhagen N, Denmark

L is the length of the junction. On the 1st ZFS ($N \cong 1$) a single soliton pendles back and forth along the junction with a velocity $\cong \bar{c}$ hence producing a periodic voltage pulse train of frequency $f_1 = \bar{c}/2L$ at both ends. On the Nth ZFS the motion of N solitons is involved and N pulses are produced within one period $1/f_1$. The detailed frequency spectrum of the voltage will depend on the distribution in time of the N pulses within the period. Analogue[4,5] and numerical studies[5] have suggested that configurations where the solitons are bunched (or congealed) play an important role in explaining the dynamics of the motion. This is supported by recent detailed measurements[7] on niobium-lead junctions and by comparison with numerical simulations[11] with parameters chosen in accordance with the experiments.

In this paper we report on the detailed numerical study and comparison with the experiments on two junctions with different length-to-Josephson-penetration-depth (L/λ_J) ratio. We demonstrate that the shunted junction (SJ) model, - which leads to a perturbed sine-Gordon equation[14] - explains the experimental observations quantitatively even down to very subtle details. We find good agreement for the current-voltage characteristic, power output, and for the shape and height of the ZFS's. On the 2nd ZFS we find two distinct modes of soliton oscillations - the perfectly symmetric mode and the bunched mode - thus confirming previous results by Erné and Parmentier. On the 3rd ZFS - the highest step considered here - two simultaneous frequencies are observed. This was also seen in the experiments. By introducing a weak external field in the numerical model we are able to reproduce the manner in which the junction responds to changes in the applied bias-current.

The outline of the paper is as follows. In section II we present the SJ model, derive the perturbed sine-Gordon equation, and give the parameters relevant to the experimental data. In section III we present the results for the longer junction ($L/\lambda_J = 6$) current biased on the 1st, 2nd, and 3rd ZFS, respectively. In section IV similar results are described for the shorter junction ($L/\lambda_J = 2$). Finally in section V we give our conclusion. Our numerical method is discussed in the Appendix.

II. THE PERTURBED SINE-GORDON EQUATION AND THE SJ MODEL

An overlap geometry Josephson tunnel junction consists of two superconducting metal layers separated by a thin insulating oxide layer of uniform thickness (t_{ox}) which is small enough to permit quantum-mechanical tunnelling of electrons. The geometry is shown in Fig. 1a. We shall restrict our attention to narrow junctions with width ($W \ll \lambda_J$) which permits us to treat the electromagnetic fields in the insulating barrier as uniform in the Y-coordinate.[12] In particular this means that the applied bias-current can be considered uniformly distributed along the entire X-axis. The orientation of the coordinate system is also shown in Fig. 1a.[13] The tunnelling supercurrent is described by the two basic Josephson equations.

$$j(X) = j_o \sin\phi \tag{2.1a}$$

$$\frac{\partial\phi}{\partial T} = \frac{2e}{\hbar}V = \frac{2\pi}{\phi_o} \tag{2.1b}$$

where T is laboratory time.

Here j(X) is the Josephson current crossing the barrier/unit length in the X-direction, j_o being the maximum value. Formally, $\phi = \phi(X,T)$ is the difference between the phases of the order parameters of the two superconductors. It is however, more convenient for our purpose to define it through the second of the Josephson equations, Eq. (2.1b). Here $V = V(X,T)$ is the voltage drop across the insulating barrier. A combination of Eq. (2.1a,b) with the Maxwell equations

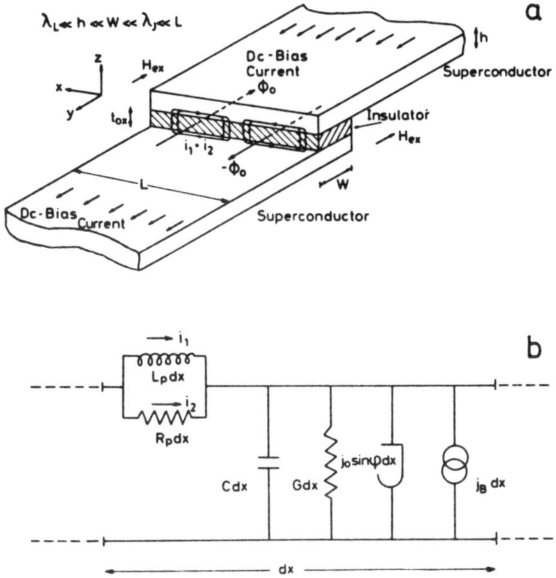

Figure 1
(a) Josephson tunnel junction of overlap geometry. (Not drawn to scale). (b) Element of lumped transmission line equivalent circuit representing the SJ model.

yields the non-linear set of partial differential equations representing the dynamics of the junction,[14,15] (c.f., Fig. 1a).

$$\frac{\partial V}{\partial X} = -i_2 R_P = -L_P \frac{\partial i_1}{\partial T} , \tag{2.2a}$$

$$\frac{\partial i_1}{\partial X} + \frac{\partial i_2}{\partial X} = -C\frac{\partial V}{\partial T} - GV - j_o \sin\phi + j_B . \tag{2.2b}$$

Equations (2.1b, 2.2a,b) also models the lumped transmission line equivalent circuit in Fig. 1b. It is this model which is often referred to as the shunted junction (SJ) model. In Eq. (2.1b, 2.2a,b), i_1 and i_2 are the superconducting and normal components of the current flowing parallel to the barrier in the X-direction. The inductance/unit length, L_P, represents the magnetic energy stored within one London penetration depth of the superconducting metal films. It is given by[16]

$$L_P = \mu_o(2\lambda_L + t_{ox})/W \tag{2.3}$$

where λ_L is the average London distance for the two superconductors (here niobium and lead), and $\mu_o = 4\pi\cdot 10^{-7}$ H/m. The series resistance/unit length, R_P, represents scattering of quasi-particles in the surface layers of the two superconductors. The capacitance/unit length, C, represents the electric energy stored in the barrier and is given by

$$C = \varepsilon_r \varepsilon_o W/t_{ox} \tag{2.4}$$

where ε_r is the relative dielectric constant of the barrier oxide layer, and $\varepsilon_o = 1/36\pi\cdot 10^{-9}$ F/m. The dissipative currents/unit length (the quasi-particle tunnelling) are represented by the term GV, where 1/G is an effective normal resistance.

The supercurrent/unit length is given by $j_o \sin\phi$. Finally j_B represents an externally applied bias-current/unit length.

Equation (2.1b, 2.2a,b) combines into the third order partial differential equation:

$$(L_P/R_P)\phi_{XXT} + \phi_{XX} - L_P C\phi_{TT} - GL_P\phi_T = (2\pi L_P j_o/\Phi_o)(\sin\phi - j_B/j_o) \tag{2.5}$$

Introduction of normalized space and time coordinates, x and t, related to laboratory coordinates by

$$x = X/\lambda_J \tag{2.6}$$

$$t = T\omega_o \tag{2.7}$$

where λ_J is the <u>Josephson length</u>

$$\lambda_J = (\Phi_o/2\pi j_o L_P)^{\frac{1}{2}} \tag{2.8}$$

and ω_o is the <u>Josephson plasma frequency</u>

$$\omega_o = (2\pi j_o/\Phi_o C)^{\frac{1}{2}} \tag{2.9}$$

yields,

$$\beta\phi_{xxt} + \phi_{xx} - \phi_{tt} - \alpha\phi_t = \sin\phi - \gamma\ . \tag{2.10}$$

Where the constants α, β, and γ are defined by

$$\alpha = \frac{G}{\omega_o C}\ , \tag{2.11}$$

$$\beta = \frac{\omega_o L_P}{R_P}\ , \tag{2.12}$$

$$\gamma = \frac{j_B}{j_o}. \tag{2.13}$$

The values of the two parameters λ_J and ω_o are especially important for the dynamics of the junction. The Josephson plasma frequency, ω_o, is the lowest possible frequency of small amplitude oscillations in the unbiased junction, whereas the Josephson length, λ_J, can be viewed as a screening length over which the magnetic field induced by the Josephson supercurrent causes noticeable changes in ϕ[15]. The propagation velocity of electromagnetic signals in the junction is $\bar{c} = \lambda_J\omega_o = (L_P C)^{-\frac{1}{2}}$. The space-time behavior of the voltages and currents is governed by Eq. (2.10), which without the dissipative terms and the external bias current is the famed sine-Gordon equation with well known soliton solutions. This fact will be used in the numerical computations presented in section III.

Using the normalized time t we rewrite Eq. (2.1b) as

$$V = (\Phi_o\omega_o/2\pi)\phi_t \equiv V_N\phi_t \tag{2.14}$$

Further we mention that Eqs. (2.6) and (2.8) combined with the time integral of Eq. (2.2a) yield

$$i_1 = -(j_o\lambda_J)\phi_x \equiv -I_N\phi_x \tag{2.15}$$

Eq. (2.15) can be used to establish the boundary conditions for the junction. The superconducting surface current, i_1, gives rise to a magnetic field perpendicular to the length direction of the junction. Thus, fixing an external magnetic field in the Y-direction (H_{ex}), at the two ends of the junction corresponds to specifying ϕ_x at these points. Considering a junction of length $\ell = L/\lambda_J$ we get

$$\phi_x(0,t) = \phi_x(\ell,t) = -\frac{H_{ex}w}{I_N} \equiv \eta \tag{2.16}$$

Here the dimensionless quantity η represents the external field. In most of our computations we have assumed "open-circuit" boundary conditions, i.e., $\eta = 0$, thus neglecting loading effects due to radiation loss. However, in order to model asymmetries in the experimental system due to small external magnetic fields, e.g., we have sometimes used values of η different from zero.

The quasi-particle tunnelling loss parameter α, is determined through Eq. (2.11) from the slope (G) of the linear part of the experimental I - V curve. Due to the nonlinear nature of this curve, α cannot be determined unambiguously. We assume α to be constant in the range $0.01 < \alpha < 0.05$ for the junctions considered here. The parameter β is determined by Eq. (2.12) which is also the inverse of the Q of the superconductor surface impedance at ω_o and describes the quasi-particle losses in the film electrodes. The value of β is readily obtainable from recent measurements.[17] Relevant junction parameters are listed in Table I.

TABLE I

Sample	63H6	65H7
L	973μm	467μm
w	119μm	67μm
ℓ	~6	~2
α	0.05	0.012
β	0.02	0.01
λ_J	$1.56\cdot10^{-4}$m	$2.25\cdot10^{-4}$m
ω_o	$5.8\cdot10^{10}s^{-1}$	$4.0\cdot10^{10}s^{-1}$

Junction parameters used in the calculations.

III. RESULTS FOR THE LONGER JUNCTION ($\ell = 6$)

The experimentally determined parameters in Table I have been used in our numerical solution of Eq. (2.10). From a Fourier analysis of $\phi_t(\ell,t)$ we obtain the dc I - V characteristic and the harmonic contents of the voltage excited at the end of the junction. These numerical results are then compared with the corresponding experimental results. From Eq. (2.14) follows that the zero order Fourier component (average value) of ϕ_t corresponds to the junction dc voltage. Thus a plot of $<\phi_t>V_N$ vs. the applied bias current $I_{dc} = j_B L$ will correspond to the experimental I - V characteristic of the junction. The higher order Fourier components of $\phi_t(\ell,t)$ will give us information on the frequencies excited.

Before we go into the detailed discussion of our results - we draw attention to the following basic experimental observation: On all the observed ZFS's in the I - V characteristic radiation with the same fundamental frequency $f_1 = \bar{c}/2L$ is found. This is somewhat unexpected, because a priori one might expect that the N soliton mode on the N'th ZFS would have the solitons arrive at either end of the junction evenly distributed in time, thus leading to a fundamental frequency of Nf_1. The experimental finding of the frequency f_1 on all the ZFS's can be accounted for by introducing the notion of congealed or bunched solitons.[7] Instead of N individual solitons, one or several bunches of solitons are formed, giving rise to an excitation with the fundamental frequency f_1. Such bunches of solitons are indeed what we observe in the numerical solutions of Eq. (2.10). The dc manifestation of our computations is shown in Fig. 2 for the first three ZFS's. The numerical result[7] is shown as a full curve. Higher order steps $(N > 3)$ have also been found experimentally, but we have not persued these numerically. Figure 2 shows good agreement between numerical and experimental results, in regard to height, shape, and position of the steps. We shall now discuss the dynamics on the individual steps.

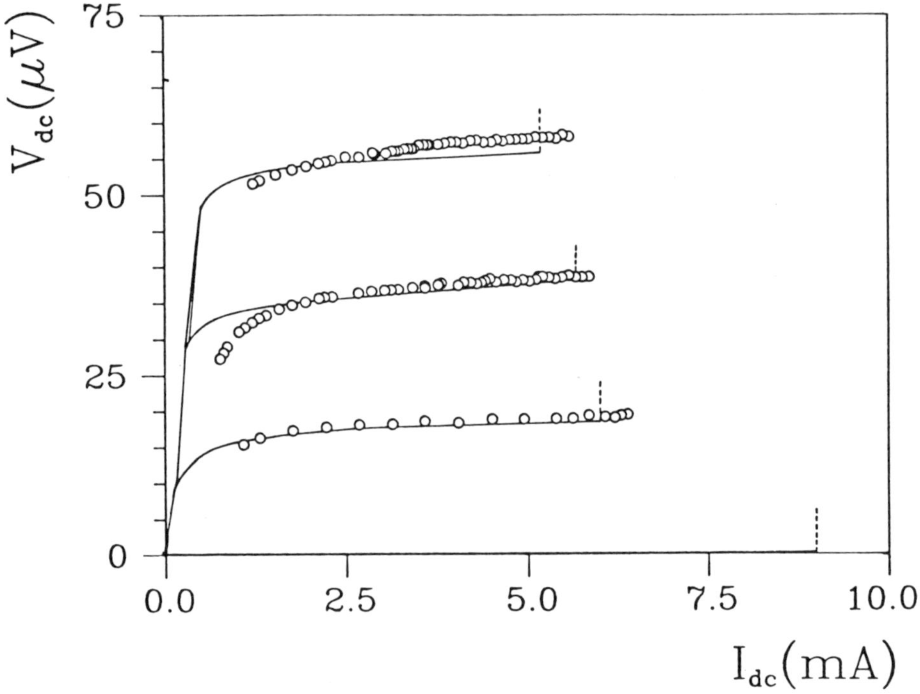

Figure 2
Dc-voltage vs. applied bias current for sample 63H6 ($\ell = 6$) showing the first three ZFS's. Circles indicate the numerical results and solid lines the experimental results.

A: 1st ZFS

In Fig. 3 we show a typical numerical result of the oscillatory motion of one soliton after steady state has been reached. (Defined operationally as the state which occurs when the first four Fourier components of $\phi_t(\ell,t)$ attained a constant value to within 2%). For runs on the 1st ZFS this state in general occurred after 1200 normalized time units. As initial conditions we have used the one-soliton solution of the unperturbed sine-Gordon equation augmented by the "ground state" solution $\phi(x,t) \equiv \sin^{-1}\gamma$, in order to prevent unnecessary generation of plasmons.[18] In order to obtain soliton confinement on the junction the soliton was launched with initial velocity equal to 0.9. The inset of

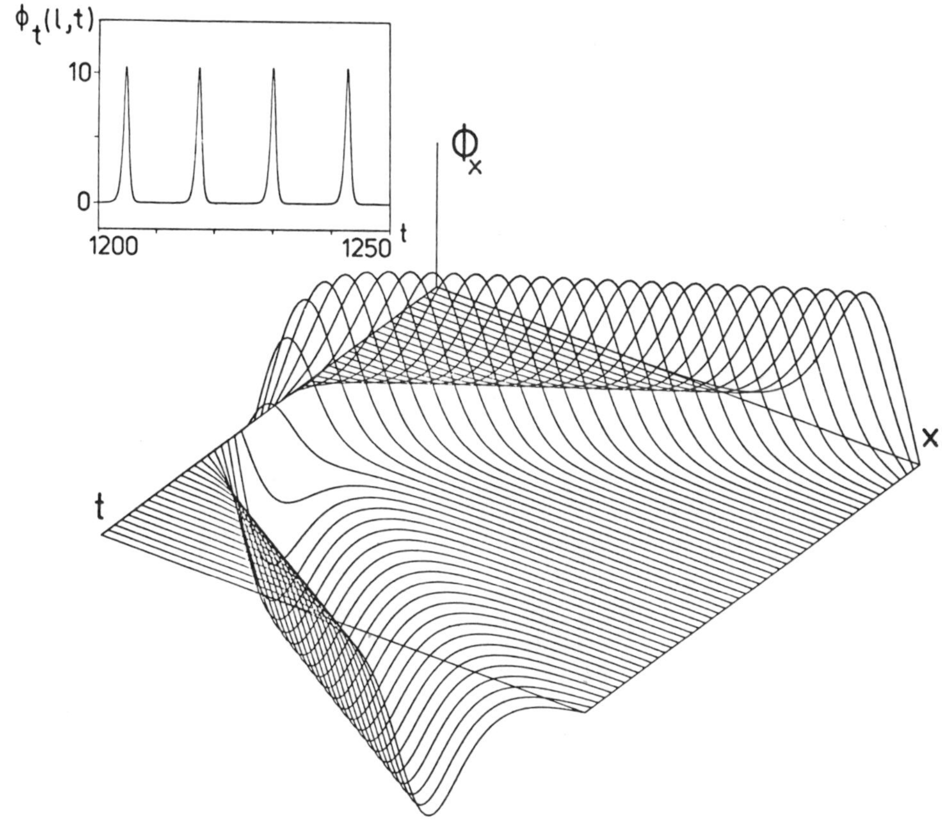

Figure 3
Numerical solution of Eq. (2.10) with $\gamma = 0.35$, $\alpha = 0.05$, $\beta = 0.02$, $\eta = 0$, and $\ell = 6$. Approximately one period of oscillation on the 1st ZFS is plotted in terms of $\phi_x(x,t)$ for 10 time units. The inset shows $\phi_t(\ell,t)$ for 50 time units.

Fig. 3 shows the voltage pulse train at $x = \ell$ after repeated reflections.

In Fig. 4 we show the harmonic contents of the power vs. the applied dc-bias current. The power is here defined as the square of the Fourier components of the voltage in arbitrary units. The oscillation is found to be stable for $0.1 < \gamma < 0.71$. For $\gamma > 0.71$ the solution disappears and the solution switches to a spatially uniform ϕ-excitation. (In the I-V curve this change corresponds to a jump from the ZFS to the ohmic background). For $\gamma < 0.1$ a transition to a static solution is found. (In the I-V curve this change corresponds to a jump from the ZFS to a point on the axis $\langle\phi_t\rangle = 0$). From Fig. 4. it is clear that the ratio of the first three harmonics is nearly constant over a wide range of bias values. However, for the higher values just below switching, the contents of all the harmonics is nearly equal. This is of course a manifestation of the fact that the soliton shape approaches the step function when velocity tends to unity. We have not observed the irregular behavior for small γ-values found by Erné and Parmentier[5] (2nd paper, Fig. 5a).

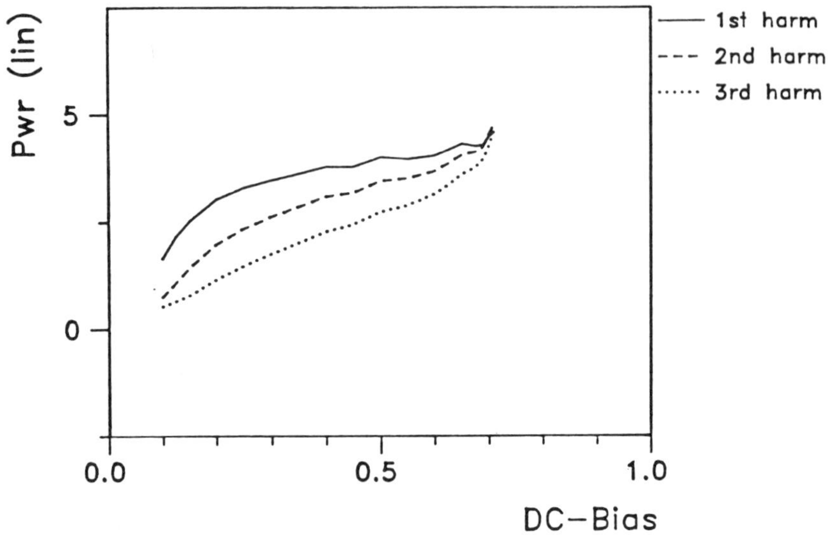

Figure 4
Harmonic contents of power vs. the dc-bias γ on the 1st ZFS. Obtained by numerical solution of Eq. (2.10) with parameters for sample 63H6.

B: 2nd ZFS

When two solitons are involved in the oscillations on the junction the picture becomes more complex. Two distinct ways of operation are now possible, - the symmetric mode and the bunched mode shown in Figs. 5 and 6 respectively. From the insets it is clear that the frequency doubles when we pass from the bunched mode to the symmetric mode. In Fig. 7 the harmonic contents of the power is shown as function of the dc-bias current. The figure was obtained in the following way: For $\gamma = 0.5$[19] we use the initial conditions consisting of two superimposed sine-Gordon one-soliton solutions (plus the $\sin^{-1}\gamma$ term) launched with equal velocity 0.9. This choice of initial conditions corresponds to a bunched mode. When steady state was obtained for ($t \cong 2000$) the computations were stopped, the bias current was increased by 0.01 (i.e. to $\gamma = 0.51$), and the computation restarted, the initial conditions now being the steady state solution obtained for $\gamma = 0.5$. This procedure was continued until a switch from the 2nd ZFS to the background curve occurred for $\gamma = 0.65$. This is the upper limit in γ for stability of the bunched mode. The process was now reversed and the bias current was decreased in steps (0.01). Continuing downwards we found transition to the symmetric mode at $\gamma = 0.16$. In Fig. 7 the 1st and 3rd harmonic vanishes for this value of γ. At $\gamma = 0.09$ the lower limit of stability was reached and the oscillations extinguished. When γ was increased again, the symmetric mode remained stable even beyond the point where the transition from bunched to symmetric mode occurred for decreasing bias. The symmetric mode finally switched from the 2nd ZFS into the solution at the background curve at the same point as the bunched mode did ($\gamma = 0.65$). Thus we have found that over a large range of γ values both the symmetric and the bunched mode are stable. Further computations for $\gamma = 0.25$ confirmed the stability of the symmetric mode as well as the bunched mode, with symmetric and bunched solitons in the initial conditions, respectively. In Fig. 8a and 8b the harmonics in the power is shown as function of time in the two cases. After a transient the harmonics settle

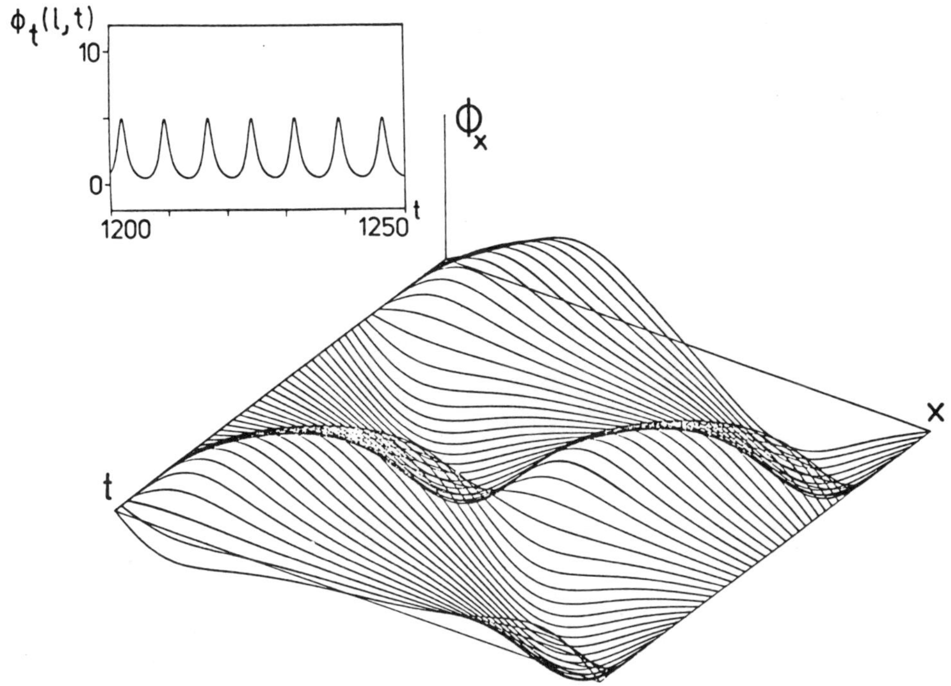

Figure 5
Numerical solution of Eq. (2.10) with $\gamma = 0.125$, $\alpha = 0.05$, $\beta = 0.02$, $\eta = 0$, and $\ell = 6$ showing the symmetric mode on the 2nd ZFS in terms of $\phi_x(x,t)$ for 10 time units. The inset shows $\phi_t(\ell,t)$ for 50 time units.

down at constant values. In Fig. 8a the 2nd harmonic is dominant, corresponding to the symmetric mode. The ripples on the 1st and 3rd harmonic 50 dB below is due to numerical noise. In Fig. 8b the bunched mode is characterized by the dominant 1st and 3rd harmonics.

We emphasize that our computational way of changing γ simulates the experimental way of regulating the bias current.[7] Thus the I-V characteristic (Fig. 2) has been obtained experimentally by careful sweeps in the bias current. The change from a bunched mode into a symmetric mode for decreasing values of γ has also been found experimentally. However, starting in a symmetric mode and increasing γ we observe no spontaneous transition to the bunched mode as is found experimentally. In the experimental set-up small asymmetries due to weak external magnetic fields cannot be completely ruled out. In our calculations we model this feature by introducing non-zero boundary conditions (e.g., $\eta = 0.1$ in Eq. 2.17).[20] In the inset in Fig. 7 the resulting transition region between bunched and symmetric modes is shown. Only a small reminiscent of the hysteresis phenomenon observed for $\eta = 0$ is left and a reversible transition from symmetric to bunched mode is found for $\gamma = 0.16$. The spontaneous transition is also shown in Fig. 9, where the effect of a gradual increase of the external magnetic field is illustrated for a fixed value of bias current ($\gamma = 0.25$). The figure was obtained in the following way: At $t = -2500$ we started a symmetric mode, i.e., two sine-Gordon solitons launched with velocities 0.9 and -0.9, respectively. For $-2500 < t < -1250$ (not shown) we kept $\eta = 0$. At $t = -1250$ the magnetic field η was increased to 0.05. The computations were then continued until $t = 0$. At $t = 0$ the magnetic field was raised to $\eta = 0.1$, and the results up to $t =$

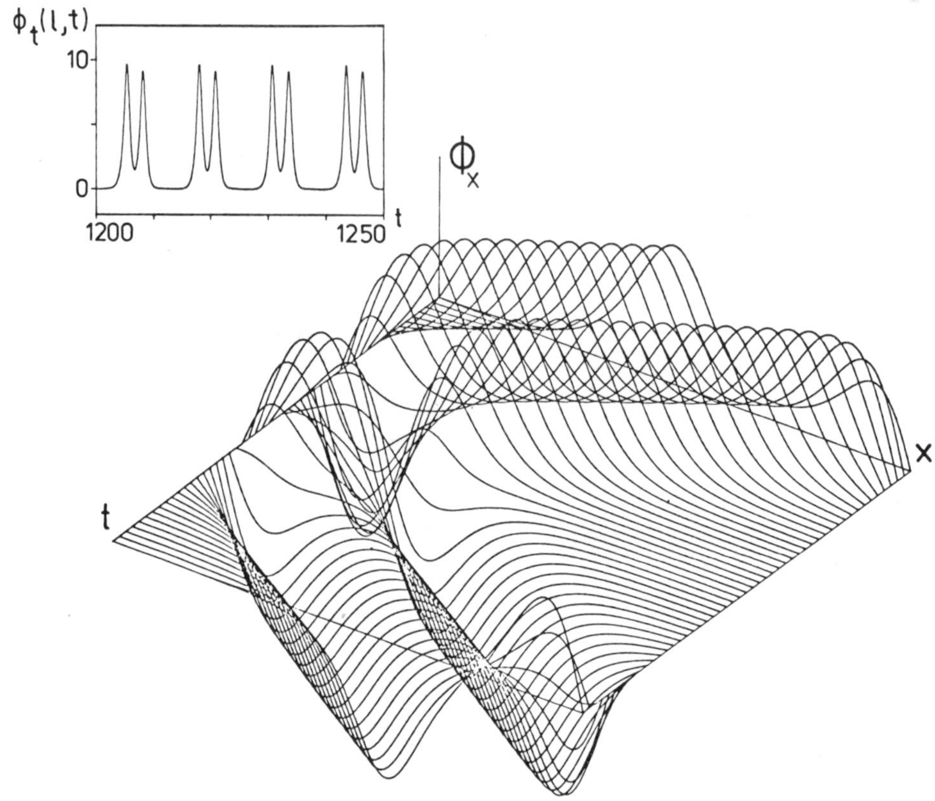

Figure 6
Numerical solution of Eq. (2.10) with $\gamma = 0.3$, $\alpha = 0.05$, $\beta = 0.02$, $\eta = 0$, and $\ell = 6$ showing the bunched mode on the 2nd ZFS in terms of $\phi_x(x,t)$ for 10 time units. The inset shows $\phi_t(\ell,t)$ for 50 time units.

1000 are shown in the figure. For $0 < t < 5000$ we see a rapid increase towards a constant value in the 1st and 3rd harmonic. In the interval $4000 < t < 6000$ the second harmonic drops 10 dB. These changes signify a transition from symmetric to bunched mode at $t \cong 5000$. The final levels of the harmonics compare with levels in Fig. 8b.

C: 3rd ZFS

Most phenomena observed on the 2nd ZFS reappear on the 3rd ZFS where three solitons are simultaneously excited on the junction. In Figs. 10 and 11 the 'symmetric' mode (3 single solitons) and the bunched mode (3 bunched solitons) are shown. We shall denote the two modes the 1+1+1 and the 0+3 configuration, respectively. In addition we find a third mode, where one soliton moves as a free entity and two solitons travel in a bunched configuration. We shall denote this mode a 1+2 configuration. The velocity of the single soliton is different from that of the 2-bunch. Figure 12 shows the situation where the single soliton and the 2-bunch have the maximum separation. At later times the picture looks much like that in Fig. 11. This 1+2 configuration is only found stable in a narrow range of γ values as seen in Fig. 13 where the 1st harmonic of the power vs. γ is shown. The 1+2 configuration manifests itself as rapid oscillations in the 1st harmonic in ther interval $0.34 < \gamma < 0.41$. Figure 13 is obtained in the

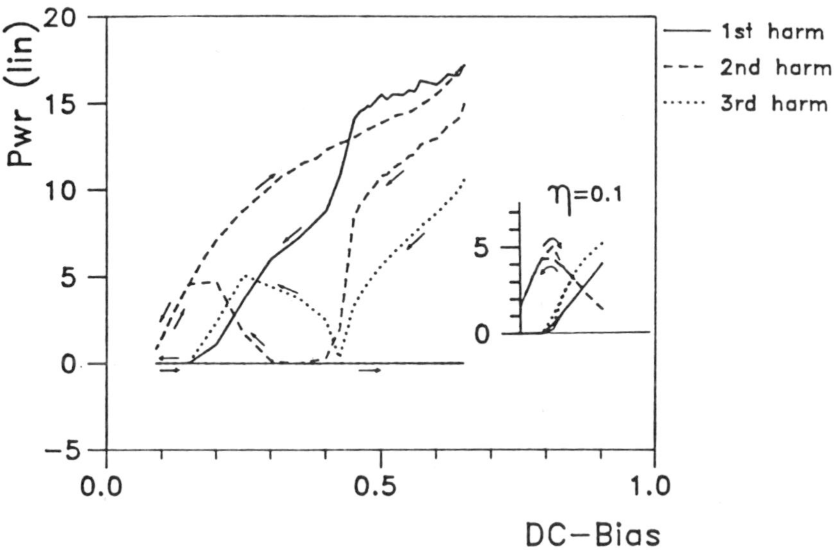

Figure 7
Harmonic contents of power vs. the dc-bias γ on the 2nd ZFS. Obtained by numerical solution of Eq. (2.10) with parameters for sample 63H6. The arrows indicate the direction in which the bias current was varied. The inset shows the near vanishing of hysteresis when a small external magnetic field is applied (η = 0.1).

same manner as Fig. 7. In Fig. 14 the 1st, 2nd and 3rd harmonics are shown as functions of time for a number of selected γ values in the range $0.34 < \gamma < 0.42$. For $\gamma = 0.34$ (Fig. 14a) and $\gamma = 0.42$ (Fig. 14d) the 1st harmonic settles to a constant value larger than the 2nd and 3rd harmonics. This implies the 0+3 configuration for these γ values. In Fig. 14b ($\gamma = 0.36$) and Fig. 14c ($\gamma = 0.40$) an oscillatory variation in the harmonics is found indicating the coexistence of oscillations at two adjacent frequencies. The 2-soliton bunch move with velocity u thus giving rise to the fundamental frequency $f_1 = u/2\ell$. The single soliton moves with velocity $(u+\Delta u)$ giving rise to a second frequency separated from f_1 by $\Delta f = \Delta u/2\ell$. This simultaneous excitation of two adjacent frequencies was also found experimentally on both the 3rd ZFS and the higher order ZFS's.[7,21] The experimentally separation observed was approximately 50 MHz. The oscillation period in Fig. 16c ($\gamma = 0.40$) indicates that $\Delta f \cong 23$ MHz in reasonable agreement with the experiments. (For this estimate Eq. (2.7) and ω_o (Table I) have been used). For lower values of γ the frequency separation, Δf, gets smaller and vanishes completely at $\gamma = 0.34$. When the bias current is decreased further, we find a transition to the symmetric mode at $\gamma = 0.25$, in a manner similar to that found on the 2nd ZFS. The symmetric mode is stable down to $\gamma = 0.15$ where the solution switches from the 3rd ZFS to the background curve. Like on the 2nd ZFS no spontaneous transition from the 'symmetric' mode (1+1+1) to a bunched mode (0+3 configuration) is found for increasing γ-values, nor does the 1+1+1 mode enter the 2+1 configuration where γ varies through the interval (0.34, 0.42). This calculation was carried no further than to $\gamma = 0.42$.

Introducing an external magnetic field ($\eta = 0.1$) did not change any qualitative aspects of the transition from the 0+3 configuration to the 1+2 configuration. In Fig. 15 the time evolution of the harmonics is shown for two cases.

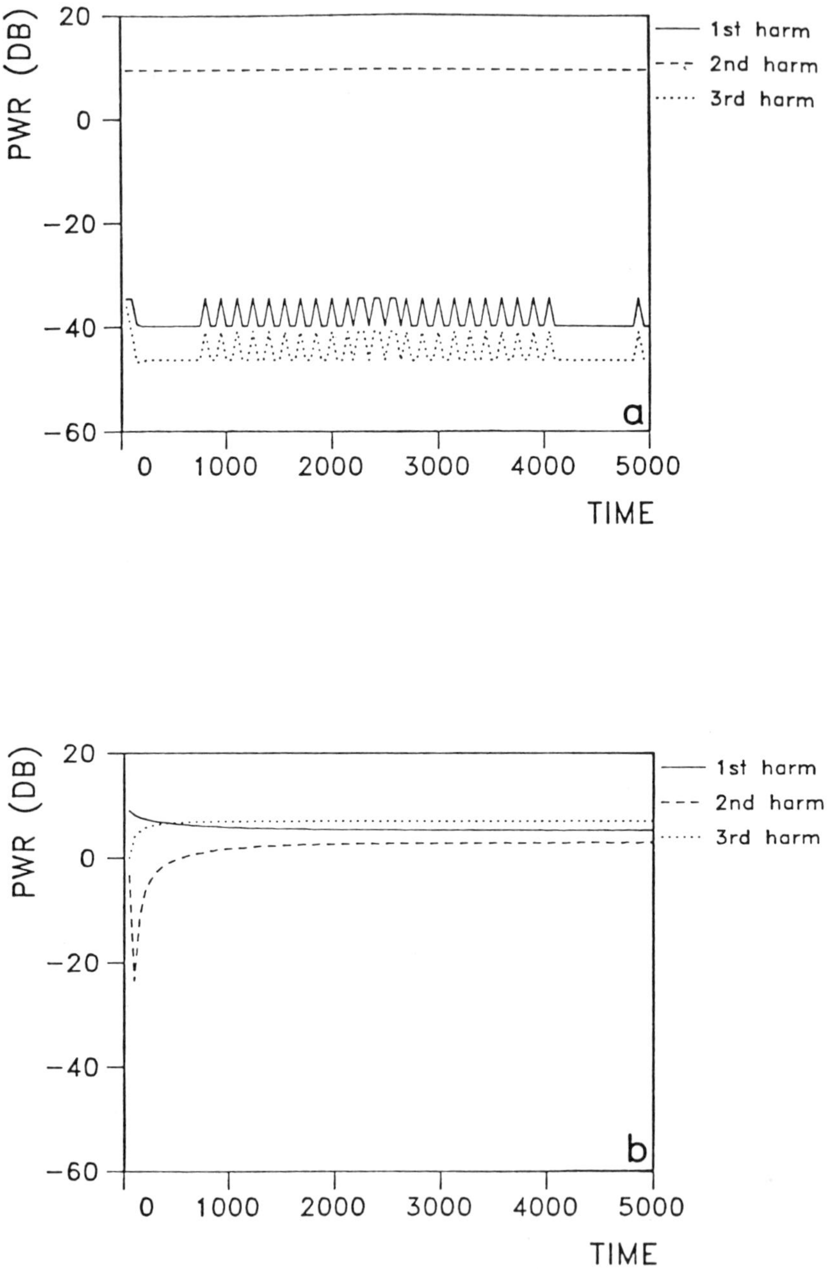

Figure 8
Harmonics in the power vs. time on the 2nd ZFS obtained by numerical solution of Eq. (2.10) with $\gamma = 0.25$, $\alpha = 0.05$, $\beta = 0.02$, $\eta = 0$, and $\ell = 6$. (a) Initial conditions representing a symmetric mode. (b) Initial conditions representing a bunched mode. Both modes are stable.

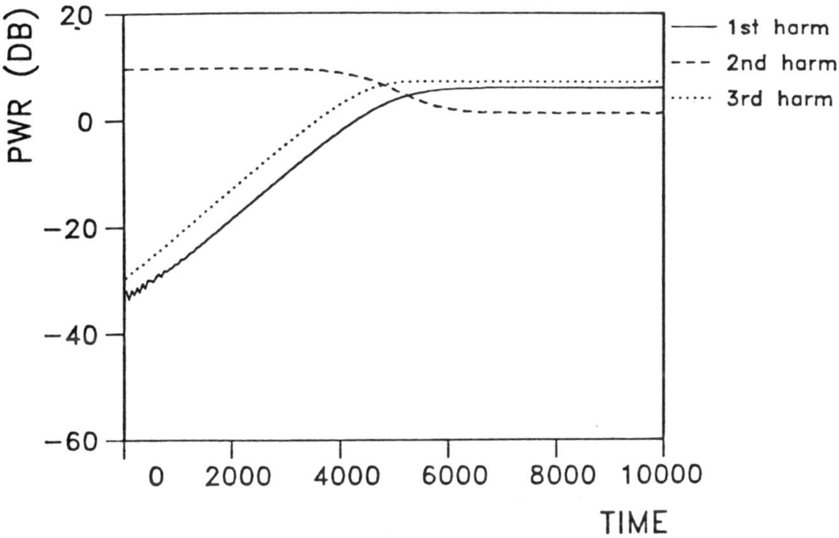

Figure 9
Transition from the symmetric to the bunched mode when an external magnetic field is gradually increased up to $\eta = 0.1$ (see text). The other parameters are as in Fig. 8.

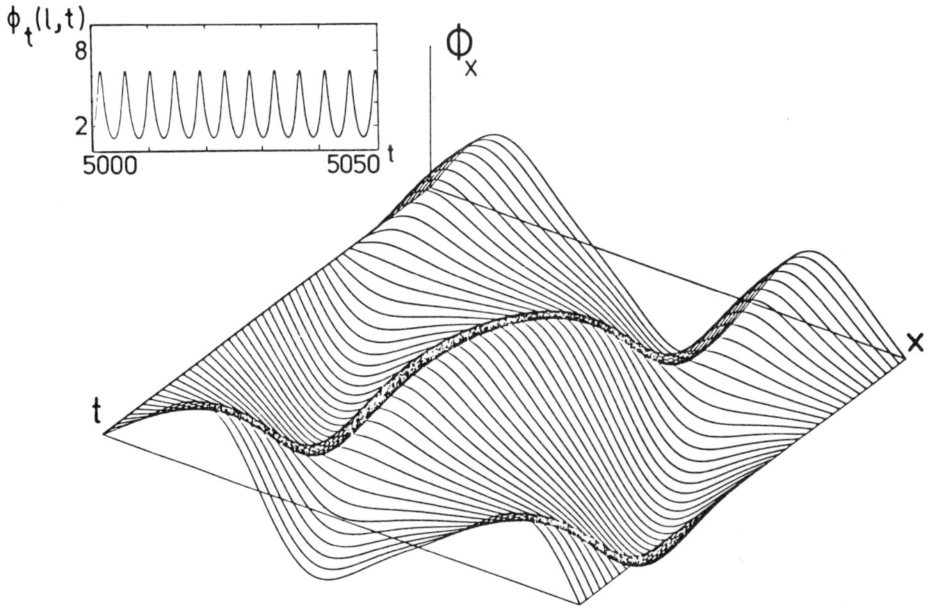

Figure 10
Numerical solution of Eq. (2.10) with $\gamma = 0.2$, $\alpha = 0.05$, $\beta = 0.02$, $\eta = 0$, and $\ell = 6$ showing the 'symmetric' mode (1+1+1) on the 3rd ZFS in terms of $\phi_x(x,t)$ for 10 time units. The inset shows $\phi_t(\ell,t)$ for 50 time units.

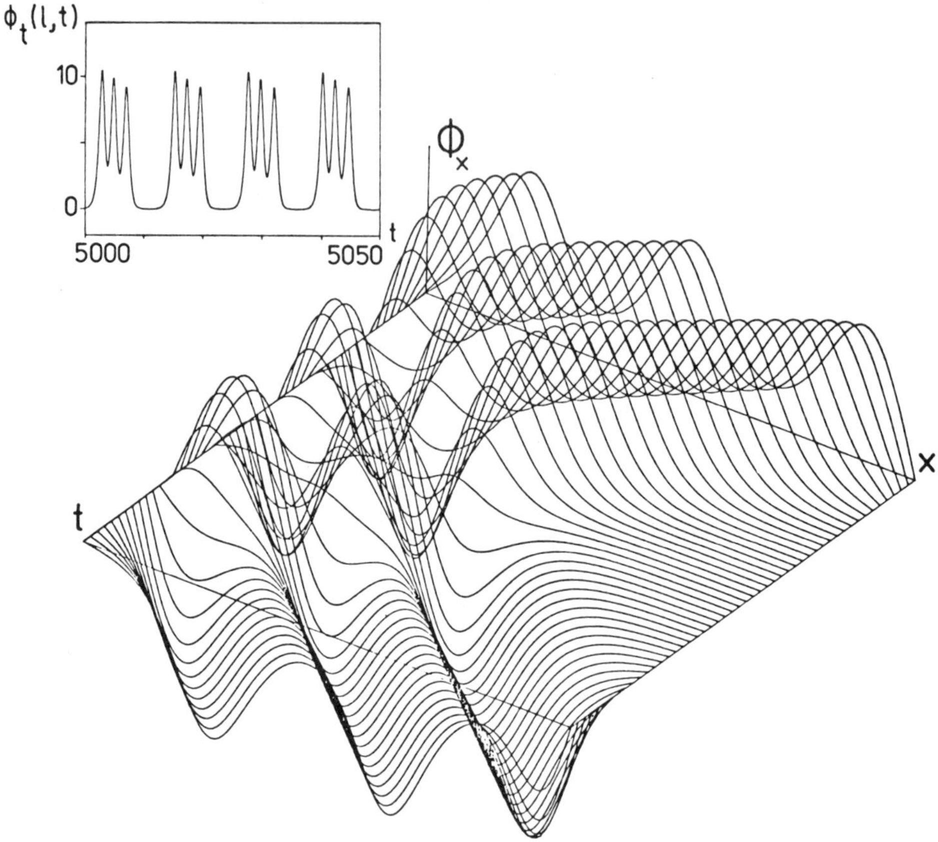

Figure 11
Numerical solution of Eq. (2.10) with $\gamma = 0.35$, $\alpha = 0.05$, $\beta = 0.02$, $\eta = 0$ and $\ell = 6$ showing the bunched mode (0+3) on the 3rd ZFS in terms of $\phi_x(x,t)$ for 10 time units. The inset shows $\phi_t(\ell,t)$ for 50 time units.

In Fig. 15a ($\gamma = 0.34$, $\eta = 0.1$) steady state is reached earlier than for $\gamma = 0.34$, and $\eta = 0$ (Fig. 14a). Also the magnitudes of the 2nd and 3rd harmonics are interchanged. The oscillatory behavior of the harmonics is Fig. 15b ($\gamma = 0.40$, $\eta = 0.1$) is also somewhat modified compared to the behavior in Fig. 14c ($\gamma = 0.40$, $\eta = 0$). We found the position and the range of stability of the 2+1 mode unchanged as a small external magnetic field was applied. Also the range of stability shows no sensitive dependence on the initial conditions as long as these have the 0+3 or the 2+1 configuration.

IV. RESULTS FOR THE SHORTER JUNCTION ($\ell = 2$)

Experimental results are also available for a function of relative short length ($\ell = 2$). The relevant parameters for this sample (65H7) are given in Table I. We have performed numerical calculations similar to those described in section III, in order to ascertain whether the soliton picture applies for shorter junctions too. One might expect the soliton confinement to be more difficult on these junctions.[22]

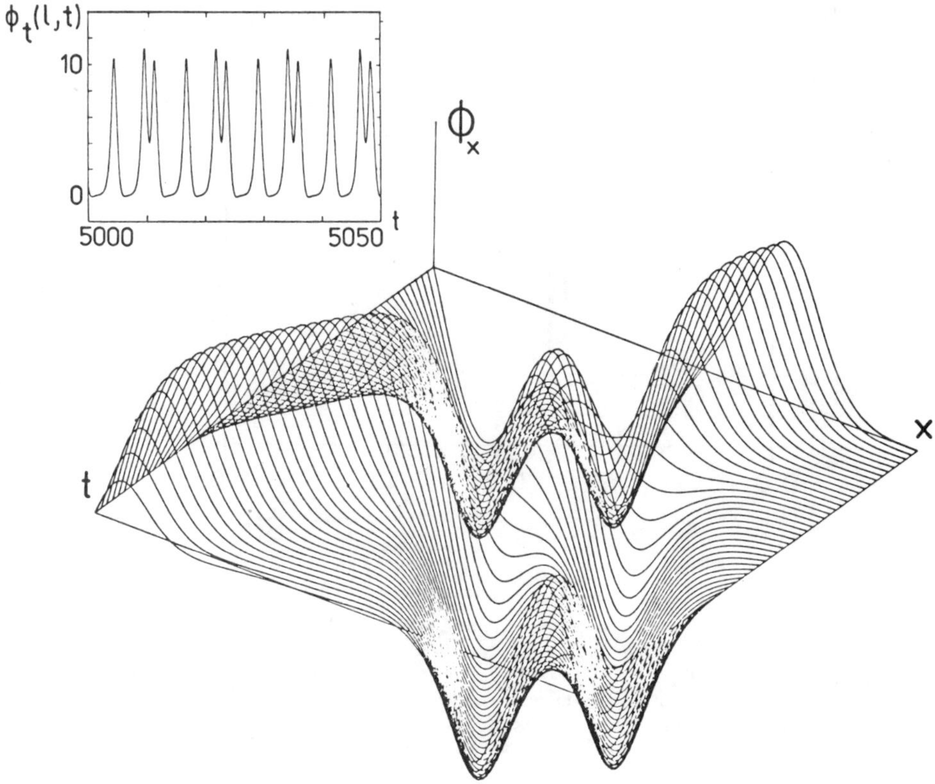

Figure 12
Numerical solution of Eq. (2.10) with $\gamma = 0.40$, $\beta = 0.02$, $\eta = 0$, and $\ell = 6$ showing the (1+2) bunched configuration on the 3rd ZFS in terms of $\phi_x(x,t)$ for 10 time units. The inset shows $\phi_t(\ell,t)$ for 50 time units.

In Fig. 16 we show the experimentally (solid curve)[7] and numerically (points) obtained I - V curve. Both in the experiments and in the computations only three ZFS's were found. There is a good agreement between the shape and position whereas the height of the computed steps is generally lower than the measured ones. Experimentally the short junction showed the same characteristic features as found on the longer junction - emission of microwave power with the same fundamental frequency on all ZFS's. Thus we expect the bunched soliton configurations to be stable also on the short junction. In Fig. 17-19 we show typical results of the oscillatory behavior on the three steps. Approximately half a period of steady state oscillation on the 1st ZFS is shown in Fig. 17 ($\gamma = 0.4$). The bunched configuration on the 2nd ZFS is shown in Fig. 18 ($\gamma = 0.3$). When γ was decreased a transition to the symmetric mode was seen for $\gamma = 0.15$. The irreversible behavior found for the longer junction is also seen here. Thus no transition from symmetric mode to bunched mode is found for increasing γ. However, introduction of a small asymmetry in the model, through the boundary conditions e.g. - again removes the hysteresis. On the 3rd ZFS only the bunched mode was found to be stable, shown in Fig. 19 for $\gamma = 0.25$.

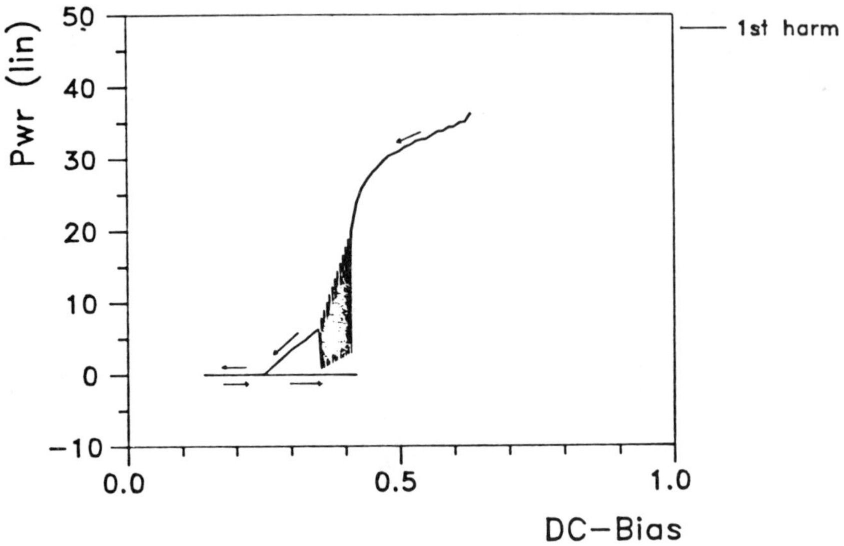

Figure 13
The 1st harmonic of the power vs. the dc-bias γ on the 3rd ZFS obtained by numerical solution of Eq. (2.10) with parameters for sample 63H5. The arrows indicate the direction in which the bias current was varied. The rapidly oscillating raange indicate the co-existence of two adjacent frequencies in the signal.

A comparison between the measured and calculated power emission at the fundamental frequency f_1 is shown in Fig. 20 vs. the bias current. The calculated output power is determined through the relation $P = V^2(f_1)/2R_L$ where we use the load resistance R_L as a fitting parameter. Using the value indicated by the arrow we find $R_L = 17$ kΩ which indicate that the open circuit boundary condition ($\eta = 0$ in Eq. 2.16) is a good approximation. This result also shows that the experimental matching to the ambient microwave circuit is bad. In Fig. 20 the solid curves are the numerical results and the points indicate the experimental observations. The agreement is good for the 1st ZFS, but gets worse for the 2nd and 3rd ZFS. However, the qualitative behavior and the relative power levels are well reproduced.

V. CONCLUDING REMARKS

By detailed numerical calculations we have shown that the SJ model - which lead to a perturbed sine-Gordon equation - is able to explain recent experimental observations on junctions of different length. We conclude that the soliton picture applies well on both relatively long ($\ell = 6$) and intermediate length ($\ell = 2$) junctions. Our calculations show the existence of the bunched soliton configuration which is crucial for the understanding of the internal dynamics of the junctions. The bunched oscillation gives rise to a signal with fundamental frequency f_1, while the 'symmetric' mode found on the lower part of the Nth ZFS yields the frequency Nf_1. We remark that even though the symmetric mode shown in Fig. 5 and Fig. 10 are solutions of the perturbed sine-Gordon equation (Eq. 2.10) it might be more natural to interpret them as current driven cavity

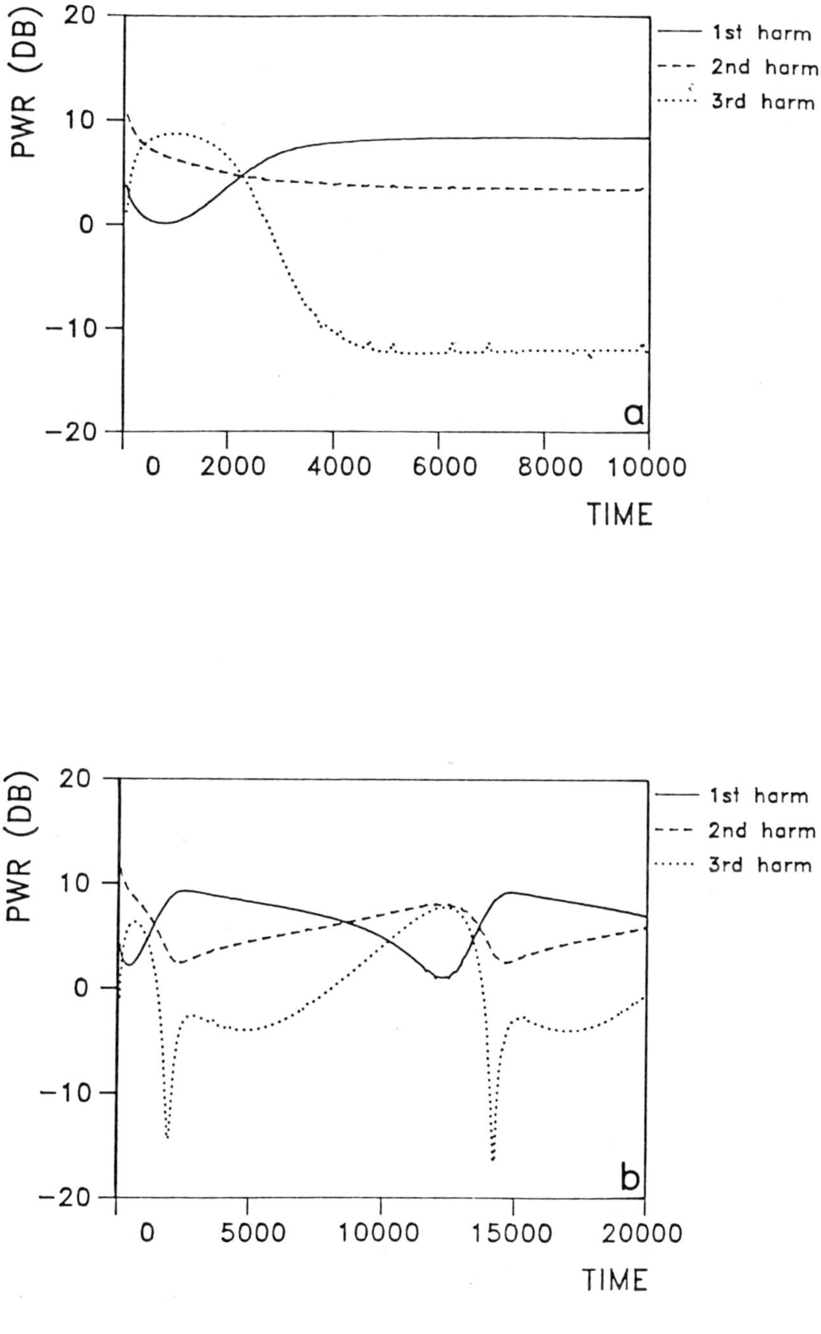

Figure 14 (a) and (b)

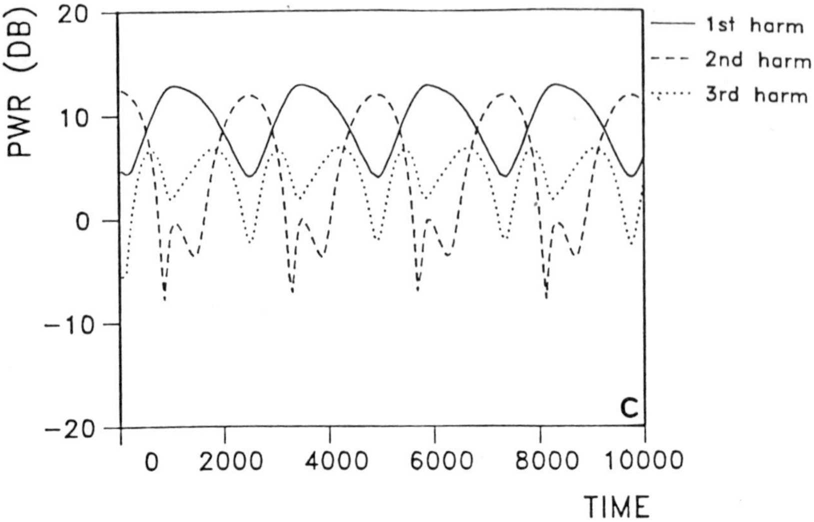

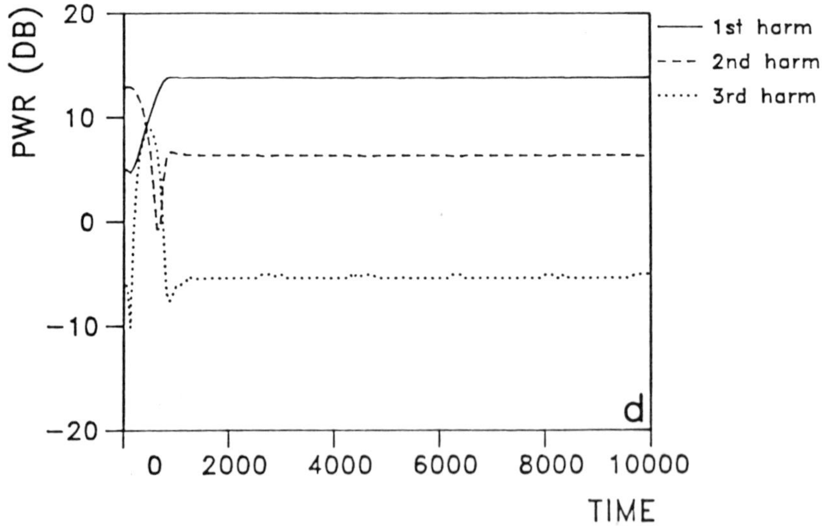

Figure 14 (c) and (d)
Harmonics in the power vs. time on 3rd ZFS obtained by numerical solution of Eq. (2.10) with $\alpha = 0.05$, $\beta = 0.02$, $\eta = 0$, $\ell = 6$, and (a) $\gamma = 0.34$, (b) $\gamma = 0.36$, (c) $\gamma = 0.40$, and (d) $\gamma = 0.42$.

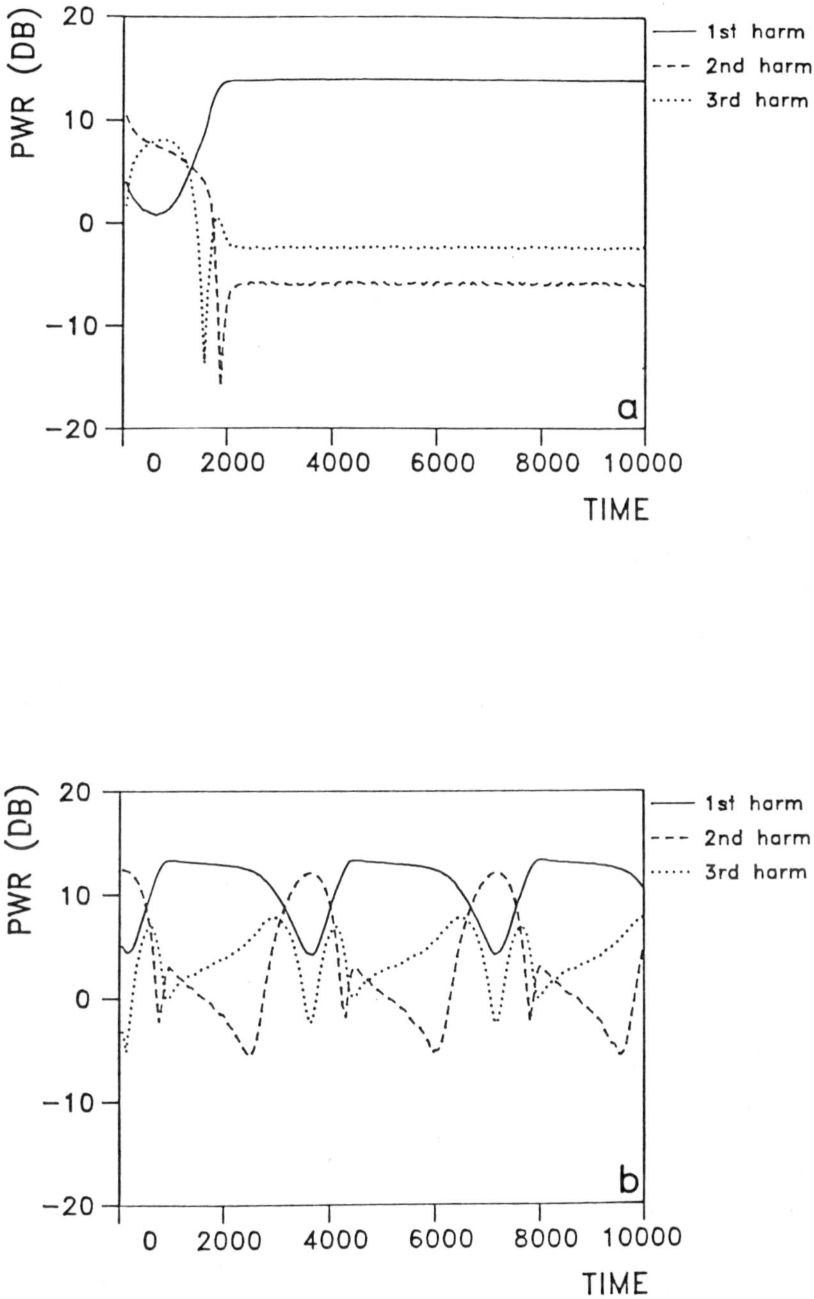

Figure 15
Harmonics in the power vs. time on 3rd ZFS obtained by numerical solution of Eq. (2.10) with $\alpha = 0.05$, $\beta = 0.02$, $\eta = 0.1$, $\ell = 6$, and (a) $\gamma = 0.34$, and (b) $\gamma = 0.40$.

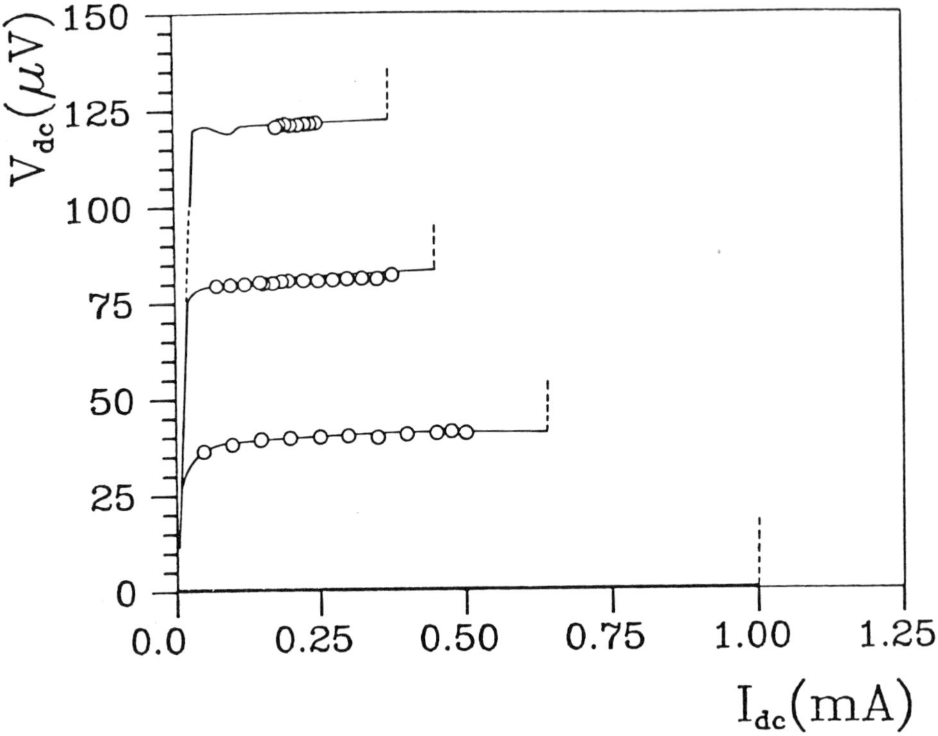

Figure 16
Dc-voltage vs. applied bias current for sample 65H7 (ℓ = 2) showing the three ZFS's. Circles indicate the numerical results and solid line the experimental results.

modes.[23] Introduction of a small asymmetry via the boundary conditions, representing a weak magnetic field e.g., yields a junction response to variations in the applied bias current in agreement with experiments. Also the bunched soliton configurations on the 3rd ZFS explain the appearance of two simultaneous signals in the experiments. Finally we mention that some work regarding the detailed effects of the loading of power from the solitons to the ambient microwave circuit still remains, also it would be interesting to see if the recently proposed[20] soliton picture for the Fiske steps applies when the calculations are followed into steady state.

ACKNOWLEDGMENTS

It is our pleasure to thank B. Dueholm, O. Levring, J. Mygind, N. F. Pedersen, and M. Cirillo very much for many useful discussions and for giving permission to use their unpublished experimental results. Also we would like to express our deep gratitude to A. C. Scott and J. C. Eilbeck who participated in the initial phase of the present study and have shown a continued interest in our work. The work was supported in part by the Danish Natural Science Research Council.

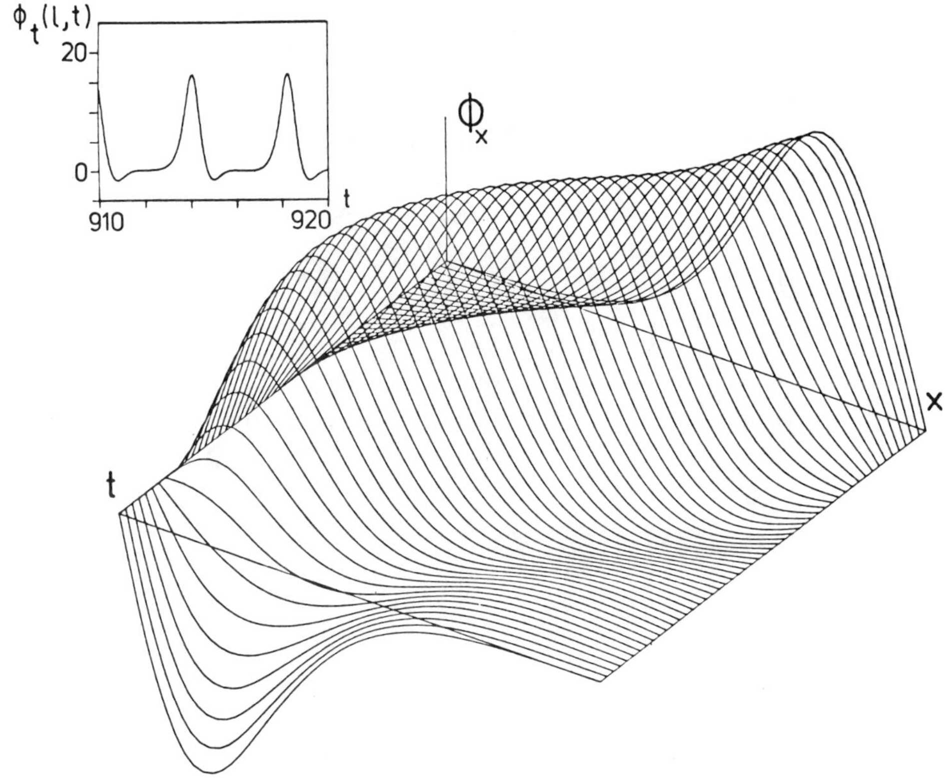

Figure 17
Numerical solution of Eq. (2.10) with $\gamma = 0.4$, $\alpha = 0.012$, $\beta = 0.01$, $\eta = 0$, and $\ell = 2$ showing approximately half a period of oscillation on the 1st ZFS plotted in terms of $\phi_x(x,t)$ for 2 time units. The inset shows $\phi_t(\ell,t)$ for 10 time units.

APPENDIX

We have used an implicit finite difference method in order to solve Eq. (2.10) numerically. Denoting the restriction of $\phi(x,t)$ to a square mesh by $\phi_i^n = \phi(ih,nk)$ we get the following approximations for the derivatives.

$$\phi_t = \frac{1}{2k}(\phi_i^{n+1}-\phi_i^{n-1}) + 0(k^2) \tag{A1}$$

$$\phi_{tt} = \frac{1}{k^2}(\phi_i^{n+1}-2\phi_i^n+\phi_i^{n-1}) + 0(k^2) \tag{A2}$$

$$\phi_{xx} = \frac{1}{2h^2}(\phi_{i+1}^{n+1}-2\phi_i^{n+1}+\phi_{i-1}^{n+1}+\phi_{i+1}^{n-1}-2\phi_i^{n-1}+\phi_{i-1}^{n-1}) + 0(h^2+k^2) \tag{A3}$$

$$\phi_{xxt} = \frac{1}{2kh^2}(\phi_{i+1}^{n+1}-2\phi_i^{n+1}+\phi_{i-1}^{n+1}-\phi_{i+1}^{n-1}+2\phi_i^{n-1}-\phi_{i-1}^{n-1}) + 0(h^2+k^2) \tag{A4}$$

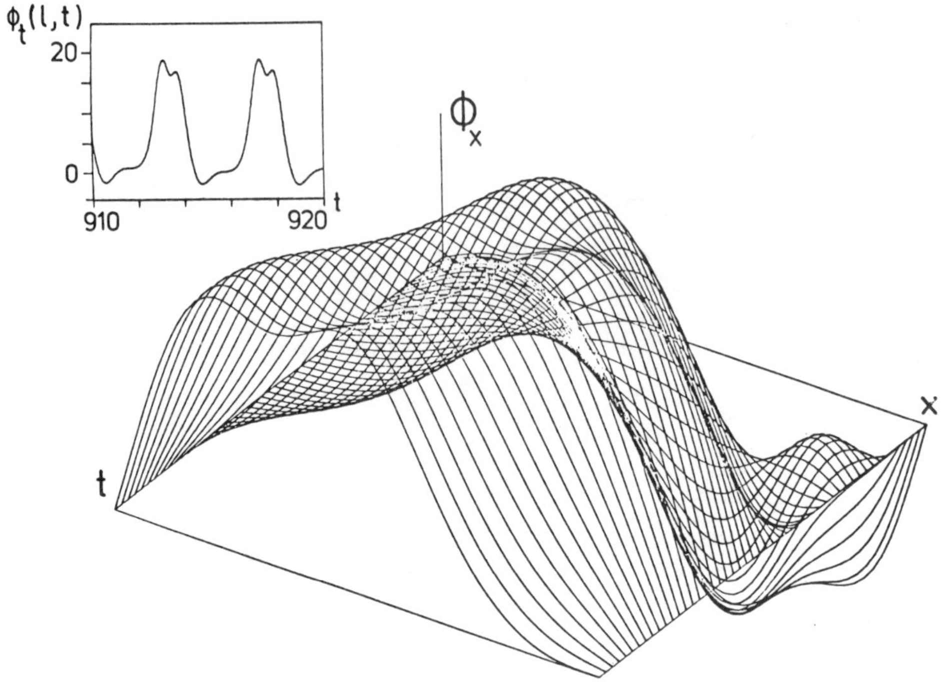

Figure 18
Numerical solution of Eq. (2.10) with $\gamma = 0.3$, $\alpha = 0.012$, $\beta = 0.01$, $\eta = 0$, and $\ell = 2$ showing the bunched mode on the 2nd ZFS in terms of $\phi_x(x,t)$ for 2 time units. The inset shows $\phi_t(\ell,t)$ for 10 time units.

Substituting Eq. (A1)-(A4) in Eq. (2.10) and neglecting terms of order $O(h^2)$ and $O(k^2)$, we get the following system of equations.

$$c_1\phi_{i+1}^{n+1} + c_2\phi_i^{n+1} + c_1\phi_{i-1}^{n+1} = c_3(\phi_{i+1}^{n-1}+\phi_{i-1}^{n-1}) - c_4\phi_i^{n-1} + c_5\phi_i^n + c_6(\sin\phi_i^n-\gamma) \ , \qquad \text{(A5)}$$

$$i = 1, 2,\ldots N \ , \quad n = 0,1,\ldots$$

where

$$c_1 = \beta+k \ , \ c_2 = -(\alpha h^2+2h^2/k+2\beta+2k) \ ,$$

$$c_3 = \beta-k \ , \ c_4 = (\alpha h^2-2h^2/k+2\beta-2k) \ ,$$

$$c_5 = -4h^2/k \ , \ c_6 = 2h^2k \ .$$

The boundary conditions are treated by introduction of imaginary points and Eq. (A5) is solved in a standard manner by means of the tridiagonal algorithm.[24] The nonlinear term in Eq. (A5) is evaluated by a predictor-corrector loop. First Eq. (A5) is used to compute the predictor $\tilde{\phi}_i^{n+1}$, this solution is then reused in Eq. (A5), but now with the nonlinear term replaced by the average $\frac{1}{2}(\sin\tilde{\phi}_i^{n+1}+\sin\phi_i^{n-1})$. The accuracy of the numerical scheme has been checked by a systematic halving and doubling of time and space steps. In the results for the longer

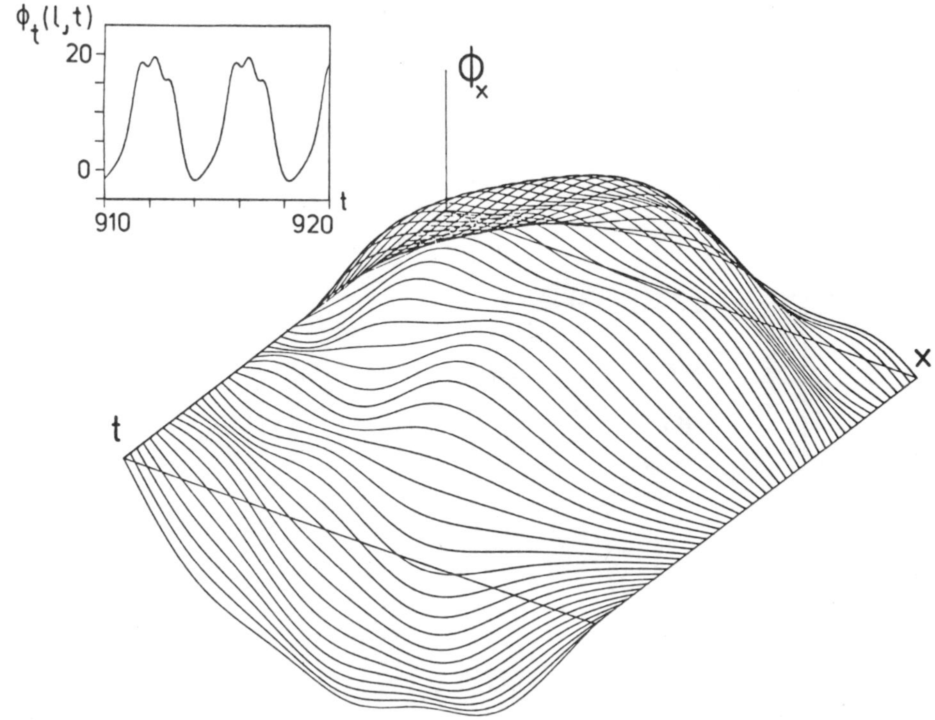

Figure 19
Numerical solution of Eq. (2.10) with $\gamma = 0.25$, $\alpha = 0.012$, $\beta = 0.01$, $\eta = 0$, and $\ell = 2$ showing the bunched mode on the 3rd ZFS in terms of $\phi_x(x,t)$ for 3 time units. The vertical scale is half of that of Fig. 17-18. The inset shows $\phi_t(\ell,t)$ for 10 time units.

junction we have used 0.025 for the timestep and 0.05 for the space step. For the shorter junction these values were 0.02 and 0.04 respectively.

REFERENCES AND FOOTNOTES

1. T. A. Fulton and R. C. Dynes, Solid State Commun. 12, 57 (1973).

2. T. A. Fulton and L. N. Dunkleberger, Rev. Phys. Appl. 9, 299 (1974).

3. K. Nakajima, Y. Onodera, T. Nakamura and R. Sato, J. Appl. Phys. 45, 4095 (1974).

4. K. Nakajima, T. Yamashita and Y. Onodera, J. Appl. Phys. 45, 3141 (1974).

5. S. N. Erné and R. D. Parmentier, Proceedings of the 1980 Applied Superconductivity Conference, IEEE, Trans, on Magnetics MAG-17, 804 (1981); J. Appl. Phys. 51, 5025 (1980); J. Appl. Phys. 52, 1091 (1981).

6. M. Cirillo, R. D. Parmentier and B. Savo, Physica 3D, 565 (1981).

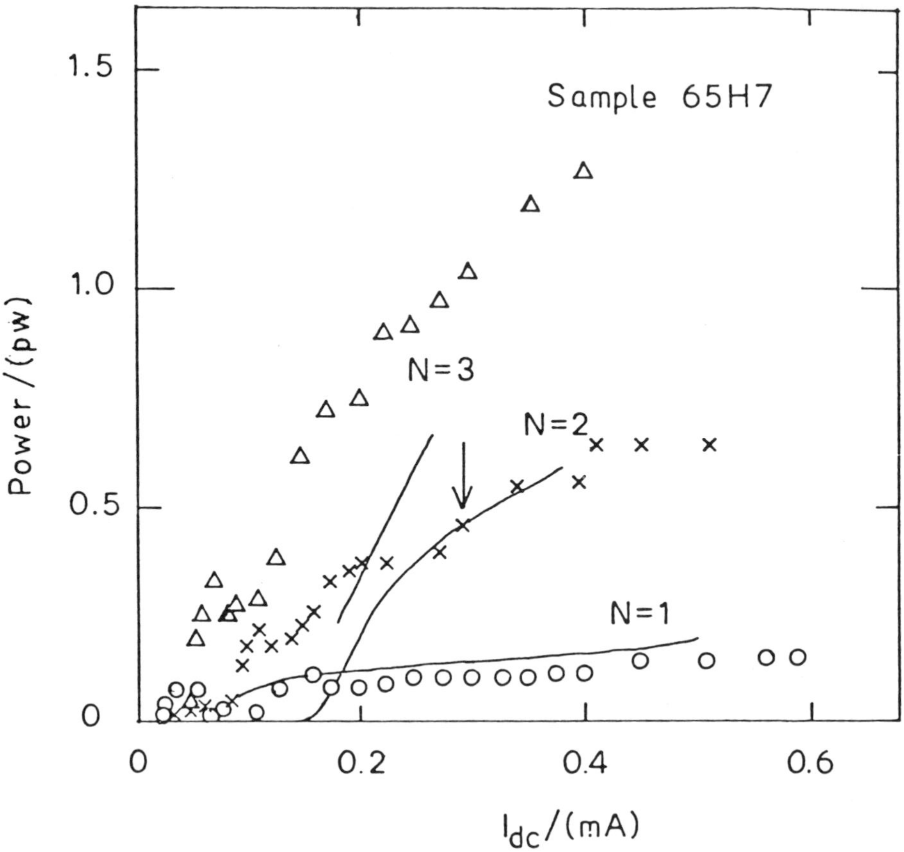

Figure 20
Microwave power emitted from the short junction ($\ell = 2$) at the fundamental frequency f_1. N = 1, 2, and 3 indicates the 1st, 2nd, and 3rd ZFS, respectively. The discrete points are experimental results and the solid lines are the numerical results. The arrow shows the power value used as a fitting parameter (see text).

7. B. Dueholm, O. A. Levring, J. Mygind, N. F. Pedersen, O. H. Soerensen and M. Cirillo, Phys. Rev. Lett. 46, 1299 (1981).

8. G. Costabile, A. M. Cucolo, S. Pace, R. D. Parmentier, B. Savo and R. Vaglio, SQUID-80, p. 147 (ed. by H. D. Hahlbohm and H. Lubbig) Walter de Gruyter, Berlin, 1980; S. N. Erné, A. Ferrigno, T. F. Finnegan and R. Vaglio. Proc. of 16th International Conference on Low Temperature Physics, Los Angeles, California, Physica 108 B+C, 1301 (1981).

9. J. T. Chen, T. F. Finnegan and D. N. Langenberg, Proc. Int. Conf. Superconductivity, August 1969, p. 413 (ed. by F. Chilton), North Holland, Amsterdam, 1971; Physica 55, 413 (1971); J. T. Chen and D. N. Langenberg, in Proceedings of the 13th International Conf. on Low Temperature Physics, Boulder, Colorado 1972, Vol. 3, p. 289 (ed. by K. D. Timmerhaus, W. J. Sullivan, and E. F. Hammel) Plenum, New York, 1974.

10. D. D. Coon and M. D. Fiske, Phys. Rev. 138, A744 (1965).

11. P. L. Christiansen, P. S. Lomdahl, A. C. Scott, O. H. Soerensen, and J. C. Eilbeck, Appl. Phys. Lett. 39, 108 (1981).

12. C. S. Owen and D. J. Scalapino, Phys. Rev. 164, 538 (1967).

13. B. D. Josephson, Advan. Phys. 14, 419 (1965).

14. A. C. Scott, F. Y. F. Chu, and S. A. Reible, J. Appl. Phys. 47, 3272 (1976).

15. T. A. Fulton, in Superconductor Applications: SQUIDS and Machines, p. 125 (ed. by B. B. Schwartz and S. Foner) Plenum, New York (1977).

16. A. C. Scott, "Solid-State Electron," 7, 137 (1964).

17. T. Yogi and J. E. Mercereau, IEEE Trans. Magn. MAG-17, 931 (1981), R. F. Broom and P. Wolf, Phys. Rev. B16, 3100 (1977).

18. An initial condition which is not an exact solution of the perturbed sine-Gordon equation will in general break up into plasmons, breathers, and kinks (antikinks). Since plasmons and breathers have $\langle \phi_t \rangle = 0$ over one period of oscillation, they cannot draw energy from the bias source and eventually they will be damped out. Choosing $\phi(x,t) \equiv \sin^{-1}\gamma$ will reduce plasmon generation because it is the correct ground state for the perturbed system.

19. This value is not very important. It is chosen only to assure that the determination of the upper stability limit for oscillations is independent of the choice of initial conditions.

20. The value $\eta = 0.1$ used here corresponds to an external field $H_{ex} \cong 1.3$ A/m. This is small compared to what is applied in the study of Fiske steps where η is of the order one. See e.g. Ref. 7 and B. Dueholm, E. Joergensen, O. A. Levring, J. Mygind, N. F. Pedersen, M. R. Samuelsen, O. H. Olsen, and M. Cirillo. Proc. of 16th International Conference on Low Temperature Physics, Los Angeles, California, Physica 108 B+C, 1303 (1981).

21. P. S. Lomdahl, O. H. Soerensen, P. L. Christiansen, A. C. Scott, and J. C. Eilbeck, Phys. Rev. B (in press).

22. O. H. Soerensen, P. S. Lomdahl and P. L. Christiansen. Proc. of 16th International Conference on Low Temperature Physics, Los Angeles, California, Physica 108 B+C, 1299 (1981).

23. J. J. Chang and J. T. Chen, Phys. Rev. B22, 2392 (1980).

24. See e.g. W. F. Ames: Numerical Methods for Partial Differential Equations, Academic Press, New York (1977), p. 52.

COUPLED NONLINEAR OSCILLATORS
J. Chandra and A.C. Scott (eds.)

NONLINEAR COUPLING OF RADIATION PULSES TO ABSORBING ANHARMONIC MOLECULAR MEDIA

Bruce W. Shore
Lawrence Livermore National Laboratory

Joseph H. Eberly
Department of Physics and Astronomy
University of Rochester

We discuss the nonlinear mutual coupling of radiation and molecules, with emphasis on the effects on radiation pulse propagation. We treat the nonlinear propagation problem numerically, with particular attention to both continuum and discrete behavior associated with the quasi-continuum model of molecular absorbers.

1. INTRODUCTION

The mutual coupling of anharmonic oscillators is one of the oldest problems in nonlinear analysis. Very recent studies of high-power laser-induced dissociation of polyatomic molecules have called further attention to various aspects of this problem. In these recent studies the laser field provides a sinusoidal force that drives the molecular oscillations into an anharmonic regime of response. However, in the laser-molecular example, the coupling is not really mutual until the molecular oscillations are able to act back on the radiation. This is precisely what does happen when the laser beam propagates through a molecular medium. In this paper we describe a series of mostly numerical analyses of the laser-molecule problem of coupled nonlinear oscillations. As we show, the most interesting new behavior (although not the only nonlinear behavior) occurs if the molecule has a so-called "quasi-continuum" band of excited states coupled to a nominal ground state.

Traditional descriptions of radiation transport in vapors assume that absorption of radiation occurs in accord with a linear Einstein rate equation. That is, the rate of change in molecular level population is assumed directly proportional to the molecular population inversion. By the same token, the absorption process is assumed to deplete the radiant intensity in accord with the linear Beer-Lambert law of attenuation: the rate of depletion of intensity with distance is locally proportional to the intensity.

As is now well-known, such traditional descriptions are valid as limiting cases, applicable to long-duration incoherent molecule-field interactions. Phenomena associated with short intense pulses of resonantly tuned and nearly monochromatic light require a more elaborate and nonlinear theoretical description. In the absence of relaxation phenomena (such as collisions), the molecular excitation is determined by the time-dependent Schrödinger equation. Thus the basic equations appropriate to coherent pulse propagation are the coupled nonlinear [1] Maxwell-Schrödinger equations (or, with the inclusion of relaxation, the Maxwell-Bloch equations [2].

The coupled Maxwell-Schrödinger equations are based on several assumptions, which we adopt. We take the interaction between the radiation field and the molecular vapor to be $-\vec{d} \cdot \vec{E}$. We assume the electric field to be a plane wave, with a variable amplitude:

$$\vec{E}(z,t) = \hat{\varepsilon} \in (t,z)\ e^{-i(\omega t - kz)} + c.c. \tag{1.1}$$

Here we will assume both the polarization vector $\hat{\varepsilon}$ and the amplitude $\in(z,t)$ to be real, for simplicity. Under the slowly-varying-amplitude approximation changes with z and t much more slowly than the carrier wave $\exp[-i(\omega t-kz)]$:

$$\frac{\partial}{\partial t} << \omega\in , \frac{\partial\in}{\partial z} << \ k\in . \tag{1.2}$$

Both of these assumptions are well satisfied by all realistic laser pulses.

The wave equation for one-dimensional propagation is

$$\left[\frac{\partial^2}{\partial z^2} - \frac{1}{c^2}\frac{\partial^2}{\partial t^2}\right]\vec{E} = \frac{4\pi}{c^2}\frac{\partial^2}{\partial t^2}\vec{P} . \tag{1.3}$$

The polarization $\vec{P}$ is defined in terms of the quantum dipole transition moments $\vec{d}_{mn}$, and the density N of the molecules:

$$\vec{P} \equiv N \sum_{E_m<E_n} \vec{d}_{mn}\, c_m^* c_n\ e^{-i(\omega t-kz)} + c.c. \tag{1.4}$$

Here the sum ranges over all of the one-photon (i.e., near-resonant) dipole-connected pairs of levels in the molecule. These are shown schematically in Fig. 1. The probability amplitudes $c_m(z,t)$ are assumed to be lowly-varying in

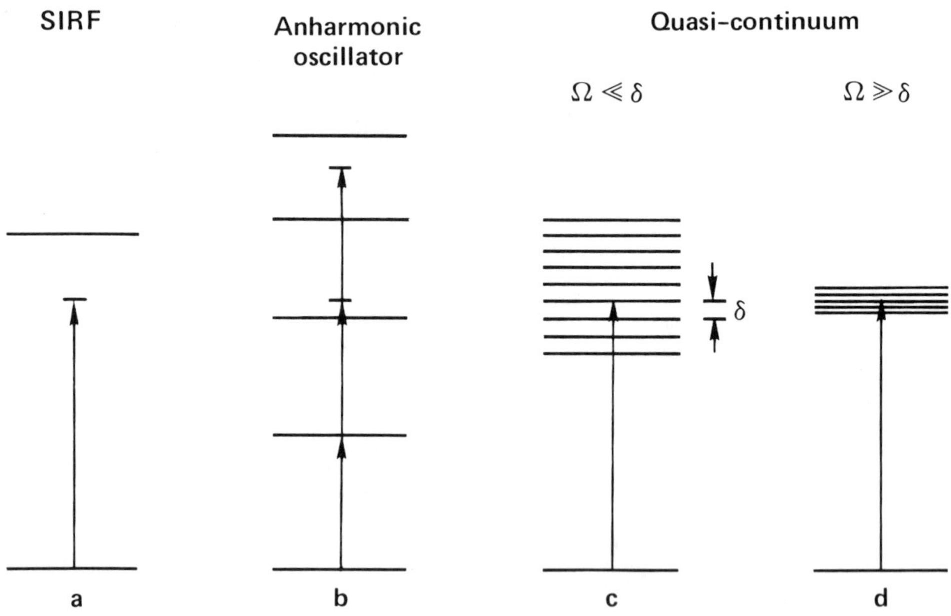

Figure 1
Models for molecular excitation a) Off-resonance two-level system for SIRF, (b) Anharmonic oscillator, (c) Quasicontinuum with level spacing much larger than Rabi frequency, and (d) Nearly degenerate quasicontinuum.

the same sense as $\in(z,t)$. Under these conditions (1.3) can be reduced to a first order equation for the electric field amplitude

$$\left[\frac{\partial}{\partial z} + \frac{1}{c}\frac{\partial}{\partial t}\right] \in = \frac{2\pi i\omega}{c} N \sum_{E_m<E_n} \hat{\in}\cdot\vec{d}_{mn} c_m^* c_n . \tag{1.5}$$

The Schrödinger equation for $c_m(z,t)$ has slightly different forms, depending on the specific molecular model being studied. For the sake of illustration, we give the form appropriate to the level structures shown in Figs. 1a, c, d:

$$\partial c_1/\partial t = -i \sum_m V_{1m} c_m \tag{1.6a}$$

$$\partial c_m/\partial t = -i \Delta_m c_m - i V_{m1} c_1 . \tag{1.6b}$$

Most of our attention will be directed to examples for which these equations are appropriate, if the index 1 refers to a nominal ground (lowest energy) state, and m refers to states in a band of higher energy and opposite-parity. The symbols Δ_m and V_{1m} refer to laser-molecule frequency detuning and laser-molecule interaction strength:

$$\Delta_m \equiv \omega_{m1} - \omega , \tag{1.7}$$

$$\hbar V_{1m} \equiv -d_{1m}\cdot\hat{\varepsilon}(z,t) . \tag{1.8}$$

The nonlinear coupling of Maxwell and Schrödinger equations is apparent in (1.5) and (1.6). The field amplitude $\in$ is determined by a sum of products of c's, and each c is determined by products of c with V (i.e., with $\in$). The extent of this mutual nonlinear coupling, in much previous work, has been reduced to the point that only one molecular transition is involved, and one speaks of having a two-level atom or molecule. In this simplest case the index m has the fixed value $m = 2$.

The Maxwell-Schrödinger equations predict a variety of fascinating and well-studied propagation phenomena including self-induced transparency, pulse reshaping, and pulse breakup into solitons [3]. These effects are associated with the simplest (two-level) model of the molecular absorber. There is reason to believe that further unusual behavior remains to be discovered as we extend our experience from the elementary two-level model to more complicated multilevel models of molecules such as shown in Fig. 1. As one begins to explore this extension, several questions come to mind:

1. How do additional (non-resonant) levels alter two level behavior?
2. Can special configurations of energy levels enhance population transfer from the ground level into the excited levels?
3. Under what conditions is coherence irrelevant?
4. What are the essential features of pulse propagation through collisionless molecular vapor?
5. What minimum configuration of levels accurately models the essence of multi-level molecular absorption.

We stress here molecular absorbers rather than atomic absorbers, having in mind the much greater complexity that accompanies the higher density of energy levels typical in polyatomic molecules.

We shall report here our observations on three models proposed as descriptive of pulse propagation through molecular media. Figure 1 diagrams the level structure of these models: (a) the Makarov-Cantrell-Louisell [4] (MCL) two-level, off-resonant propagation, which they termed Stimulated Inelastic Resonance Fluorescence (SIRF); (b) the multilevel anharmonic oscillator; and (c)-(d) the multilevel quasicontinuum.

In examining the propagation of light through a medium of multilevel systems we have benefited from numerous studies [5] of optical excitation of thin samples -- in effect the coherent response of a single multilevel molecule produced by a given pulse. This background is useful to recall as one examines the nonlinear reaction of the pulse to the induced excitation.

2. THE MCF MODEL

Soon after we had constructed the necessary computer codes to treat plane wave propagation, we read an interesting suggestion by Makarov, Cantrell and Louisell [4] pertaining to pulsed molecular excitation. They pointed out that a very abruptly initiated (square) pulse, when tuned far away from resonance in a two-level atom, would produce appreciable population excitation even far off resonance.

We examined this two-level model, as well as several-level generalizations of it. We found [6] that, although the square pulse produces the expected response, propagation diminishes the relative importance of the Bohr satellite field. Furthermore, the action of a weak resonant field in the presence of a strong nonresonant field is quite different from the action of monochromatic resonant radiation; excitation enhancement does not occur. Finally, we noted [6] that the existence of the sharply defined triplet of response functions depends critically upon the rise time of the pulse leading edge. When the pulse rises smoothly, the off-resonant excitation tends to follow it adiabatically. The molecular probability returns to the ground state after the pulse passes by.

3. PROPAGATION EXCITATION OF ANHARMONIC OSCILLATORS

The MCL paper [4] (subsequently extended in further publications,[7]) first raised the possibility that propagation modifications of square pulses could, by generating dressed-atom sidebands, provide a mechanism for exciting more population into nonresonant excited states than would be predicted for an optically thin medium. Although Eberly et. al. [6] and Macke, et. al. [8] have refuted this mechanism, a recent paper by Berman and Zaslavsky [9] raised the possibility that a temporally sharp (square pulse) plane wave, propagating through a medium of anharmonic oscillators, might become sufficiently modified by induced dipoles that appreciable excitation to high lying levels might occur, whereas anharmonicity would prevent excitation of an optically thin sample.

We tested this hypothesis by numerically modelling plane wave propagation of a square pulse through a medium of stationary (Doppler-free) 6-level anharmonic oscillators. We took the pulse carrier frequency to equal the fundamental resonance frequency of the oscillator, and took the anharmonicity sufficiently large that, in the optically thin case, the probability of populating level 6 was less than 10^{-5}. We examined a 6π pulse (that is, the Rabi area for the first transition was 6π), and we used a total time window twice the pulse duration. Figure 2 shows the pulse shape and population histories for the optically thin case with these parameters. This case is well understood. During the pulse interval the populations oscillate in a quasi-periodic manner (the system is not exactly periodic) with the ground state retaining, on average, the majority of the population. Although levels 2 and 3 each receive appreciable population, anharmonicity prevents significant population flow into levels 5 or 6; see Fig. 3. Note that the pulse leaves population in excited levels 2 and 3 (accidentally equal); this population will produce a long-time ringing after the main pulse has passed.

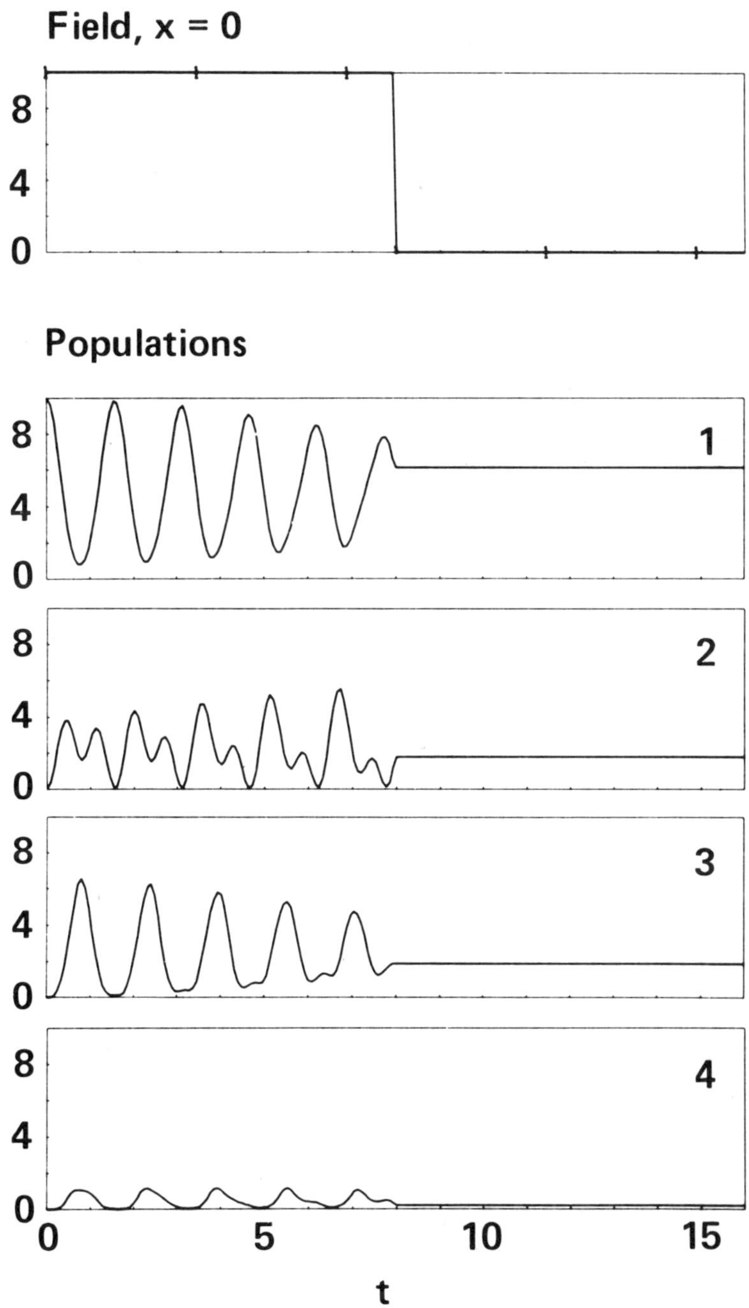

Figure 2
Optically thin square pulse; Top section: pulse intensity vs. time; Lower section: population histories.

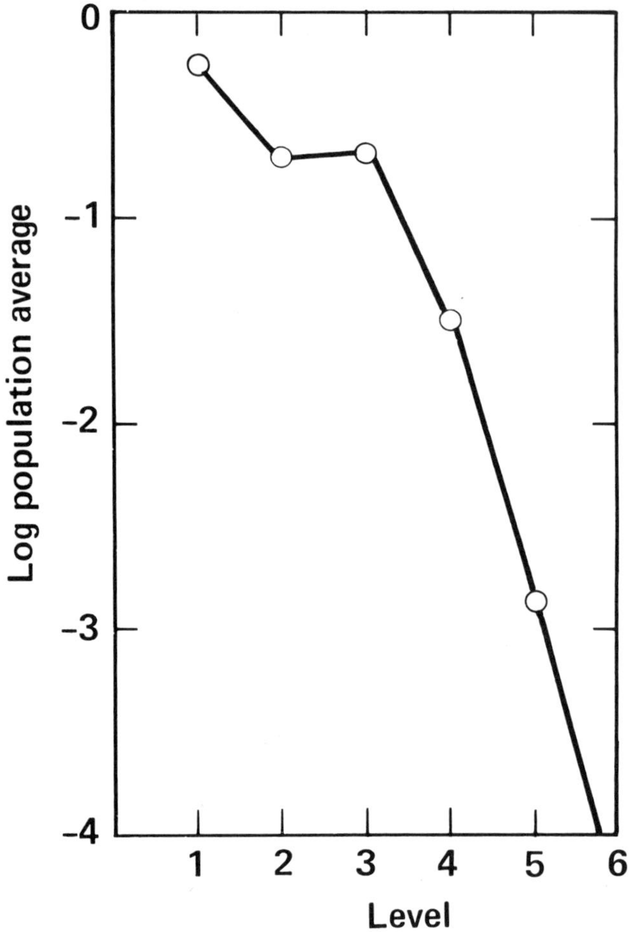

Figure 3
Distribution of time averaged populations for Fig. 1 (averaged over time window displayed).

As the pulse propagates, induced oscillating dipole moments alter the field. At first, for a short distance into the medium, the quasi-osillatory population alters the field by imposing a corresponding quasi-oscillatory modulation upon the square pulse. Figure 4 illustrates this. We also see in this figure the appearance (at around t = 10) of a low amplitude population ringing. With further propagation the pulse intensity evolves into a train of four well defined stable pulses. Figure 5 shows the pulse configuration. These pulses will continue propagating steadily, and so will eventually move off the time frame to the right.

Examination of population histories reveals no appreciable increase in high-level excitation during the course of propagation. Figure 6 plots time-averaged

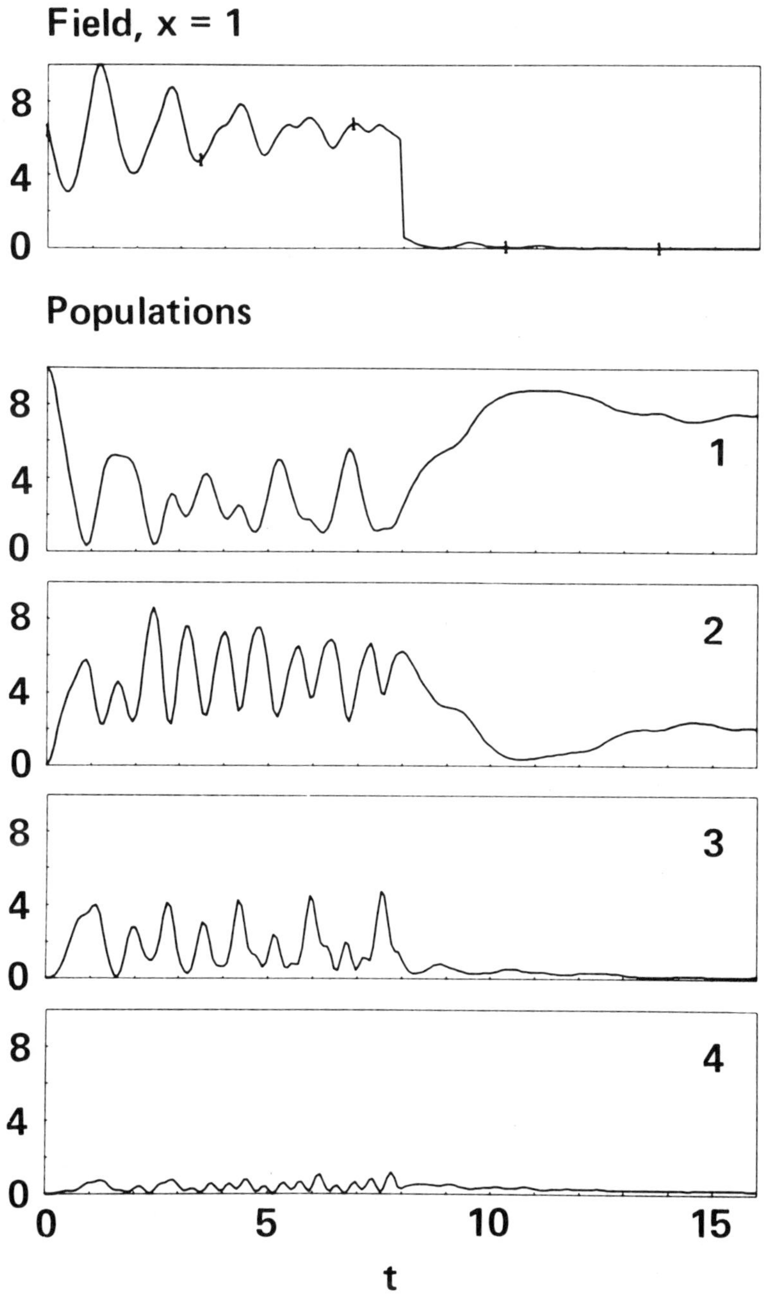

Figure 4
As in Fig. 1 after propagation to x = 1, where x is the Icsevgi-Lamb path parameter.

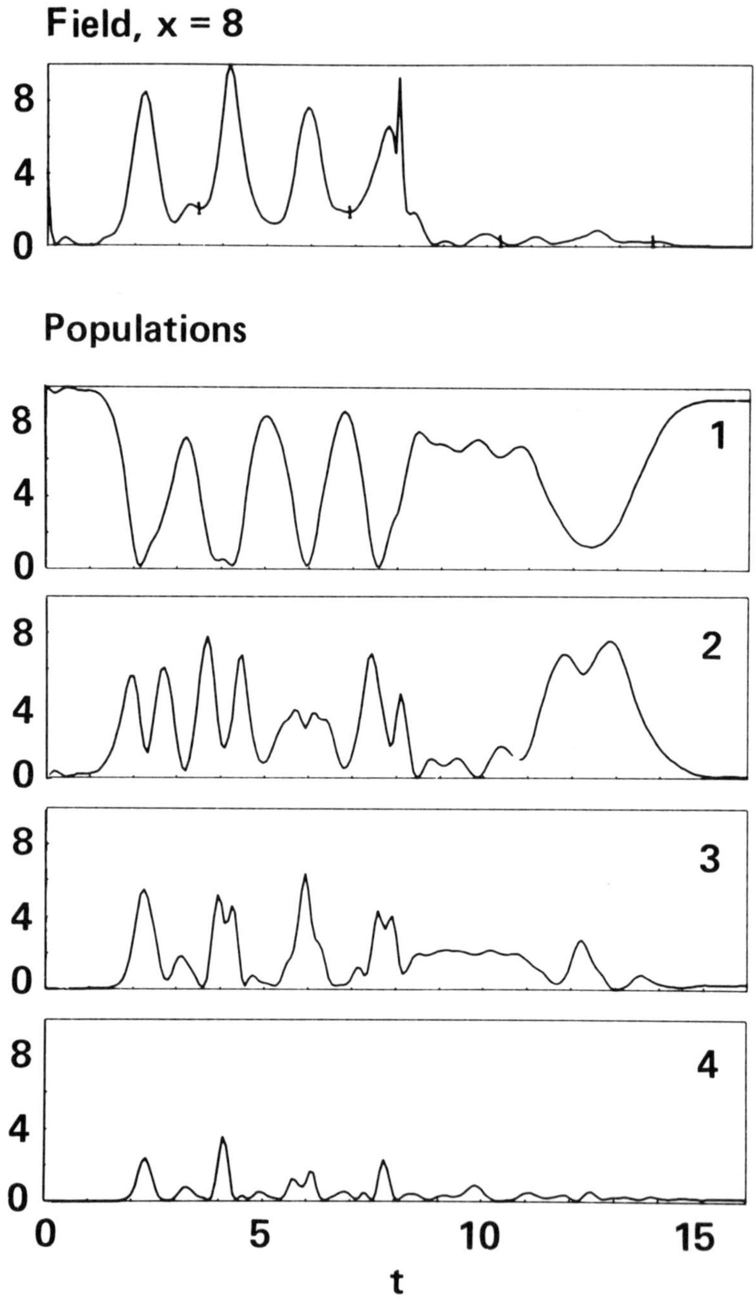

Figure 5
As in Fig. 1 after propagating to x = 8.

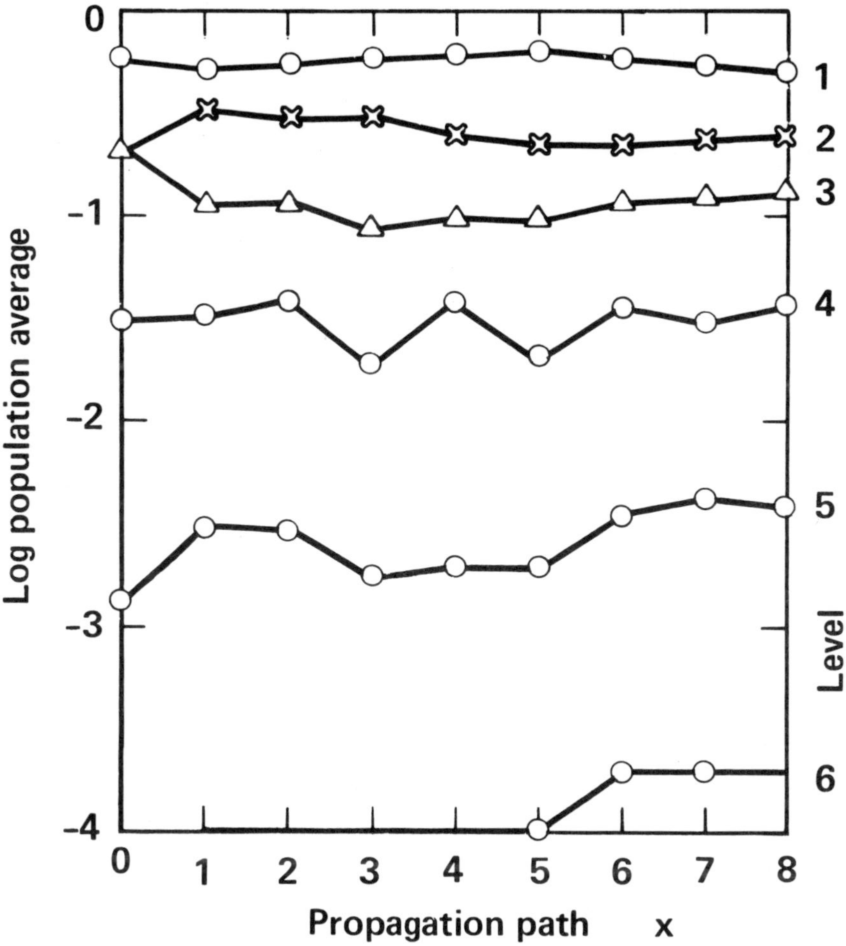

Figure 6
Time averaged populations at various values of x.

populations for a sequence of propagation distances. We see that, although the pulse has evolved into a steady-state form, the populations have undergone little change. (The miniscule populations in levels 5 and 6 do change by factors of 2 or 3, but this modest increase of small numbers does not produce significant excitation.)

We conclude that over a distance of several absorption lengths there is no evidence for propagation enhancement of excited populations, as might occur if the presence of anharmonic detuning produced a frequency chirp. The pulse itself behaves much as we would expect for a two-level atom.

4. EXCITATION OF A MEDIUM OF QUASICONTINUUM MOLECULAR RESONANCES

In many discussions of laser-induced molecular excitation [10] it is recognized that in polyatomic molecules the energy levels become increasingly dense, even for low energies, so that the molecule can behave toward photon absorption as though the molecular energy levels formed a nearly continuous distribution--the so-called quasicontinuum (QC). Because the quasicontinuum is so often invoked to explain features of laser-induced molecular excitation, it is important to understand both the underlying physics of optically thin excitations of QC as well as the QC effect on propagation. This section describes numerical studies aimed at clarifying the nature of optically thin excitation of quasicontinua.

Figure 7 shows the simplified QC model we considered [10]: an N-level quasicontinuum (N = 7 here) whose nondegenerate equally-spaced levels are all linked by equal dipole moments to the ground level. A pulsed laser with carrier frequency tuned to the central level of the QC (level 2) excites the molecule.

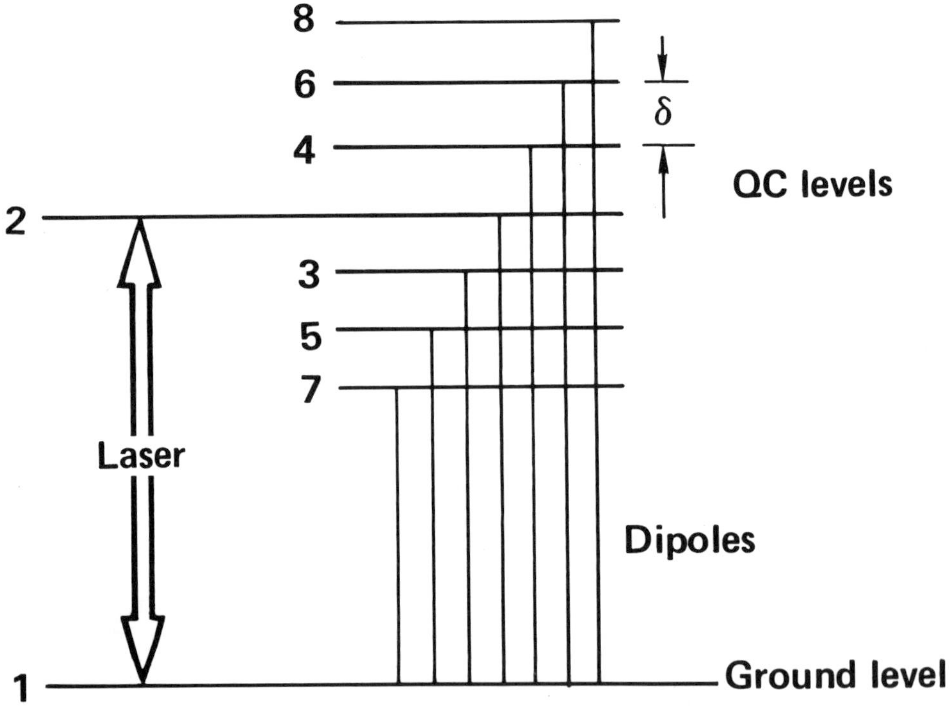

Figure 7
Diagram of 7-level quasi-continuum model showing dipole linkages to ground level 1 and laser resonance with middle (level 2) of quasi-continuum.

We let the QC levels be evenly spaced over an angular frequency band of width Γ, so that the spacing between QC levels is

$$\delta = \Gamma/(N-1) \tag{4.1}$$

and we let the dipole moments d_{1n} between ground and QC levels all be equal, so that the Rotating Wave Approximation (RWA) Hamiltonian has the form

$$[H] = \begin{array}{c|cccccc} & 1 & 2 & 3 & 4 & 5 & 6 \\ \hline 1 & 0 & v & v & v & v & v \\ 2 & v & 0 & 0 & 0 & 0 & 0 \\ 3 & v & 0 & -\delta & 0 & 0 & 0 \\ 4 & v & 0 & 0 & +\delta & 0 & 0 \\ 5 & v & 0 & 0 & 0 & -2\delta & 0 \\ 6 & & 0 & 0 & 0 & 0 & +2\delta \\ \cdot & & & & & & \\ \cdot & & & & & & \\ \cdot & & & & & & \end{array}$$

Here the interaction v (an angular frequency) is half the two-level Rabi frequency $\in d_{1n}$ for transition between the ground level and any one of the QC levels.

When the field is sufficiently weak ($v \ll \delta$), then only level 2 ever receives appreciable excitation. The population oscillates sinusoidally between level 1 and level 2 with the <u>two-level Rabi period</u> [11]

$$t_R = \frac{\pi}{v} . \tag{4.2}$$

The remaining levels, being far off resonance, do not noticeably participate in the excitation. We can therefore, in first approximation, treat these virtual levels as undergoing <u>independent</u> low-amplitude off-resonant sinusoidal population modulations. Thus the levels nearest to resonance (levels 3 and 4 in Fig. 7) oscillate at the detuned period

$$\tau = \frac{2\pi}{\delta}, \tag{4.3}$$

the next nearest levels (5 and 6 in Fig. 7) oscillate with period $\tau/2$, and subsequent levels oscillate at $\tau/3$, $\tau/4$, ... etc.

We see that τ constitutes a basic <u>recurrence time</u> [10]. For weak fields the recurrence time is much shorter than the Rabi period. Figure 8A, showing plots of level occupation probability P_n for an N = 25-level QC, illustrates this situation: we see here the sinusoidal Rabi oscillations between levels 1 and 2 as well as the higher-frequency lower-amplitude oscillations into virtual levels 3 and 4. The long-term (infinite time) population averages P_n displayed in Fig. 8B show that the virtual levels $n > 2$ have populations some 10^{-3} smaller than the resonant levels 1 and 2.

At the opposite extreme, when the field is sufficiently strong ($v \gg \delta$), we can neglect the presence of detuning and treat the system as a degenerate two-level system (see Fig. 9), having the RWA Hamiltonian matrix

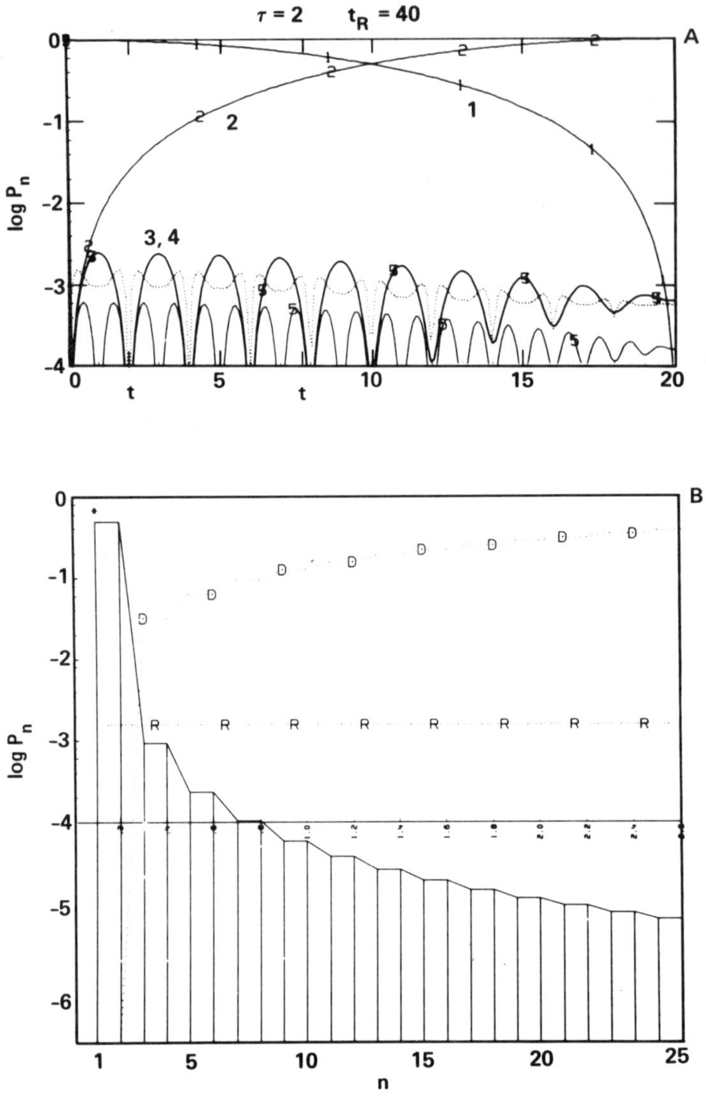

Figure 8
A) Time history of excitation, showing excitation probability P_n versus t for levels n = 1 through n = 6 of N = 26-level system. Levels 3, 4 are superposed as are levels 5, 6 (unlabeled). Dotted line shows population not in levels 1-6. Parameters are: $\tau = 2$, $t_R = 40$.

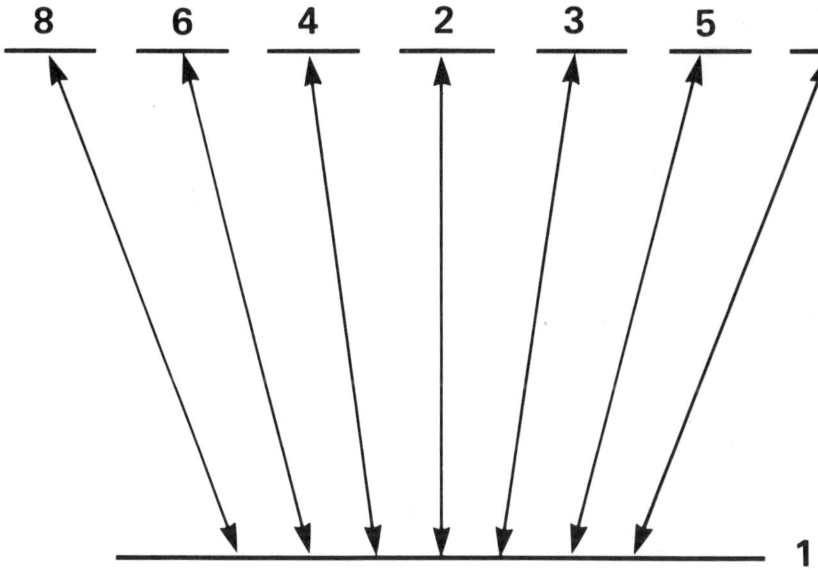

Figure 9
Linkage pattern for N = 7 level quasicontinuum in the limit of degenerage two-level system.

$$H' = \begin{bmatrix} 0 & V \\ V & 0 \end{bmatrix}$$

Here the interaction V is the root sum square of the individual interactions v:

$$V^2 = \sum_n v^2 = Nv^2 \tag{4.4}$$

and the system exhibits periodic depletion of level 1 at the band Rabi period

$$T_R = \frac{\pi}{V} = \frac{t_R}{\sqrt{N}} \ . \tag{4.5}$$

For such two-level behavior to appear, the bandwidth Γ of the QC levels must be much less than the interaction strength: $\Gamma \ll V$. Figure 10A, showing the behavior of populations under this condition, exhibits the expected oscillatory pattern of level 1 populations. As Fig. 10B shows, the long time population average distributes excitation evenly amongst the N excited levels.

In neither of these two extremes shown in Figs. 8 and 10 is there any obvious behavior indicative of a quasi-continuous distribution of energy levels. The QC becomes evident for intermediate situations, which we next examine.

The condition for weak field excitation ($v \ll \delta$) is equivalent to the requirement that the recurrence time be much shorter than the two-level Rabi period,

$$\tau \ll t_R \tag{4.6}$$

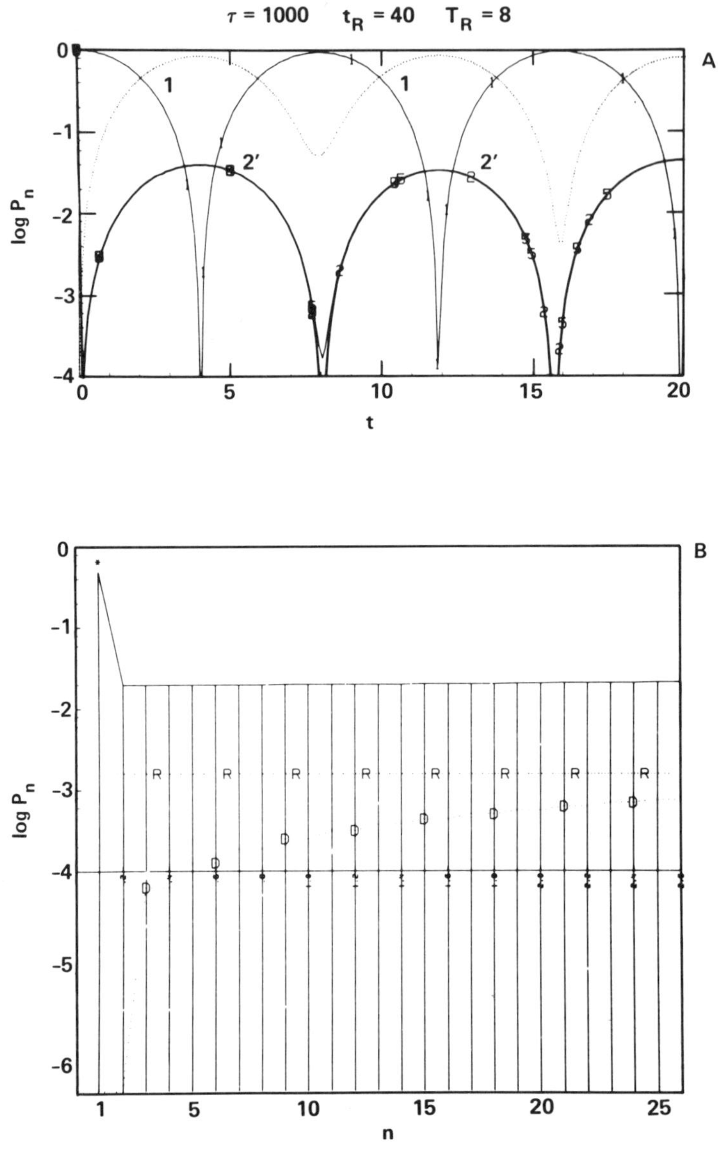

Figure 10
As Fig. 8 but with parameters τ = 1000, t_R = 40. The Band-Rabi period is $T_R = t_R/25 = 8$.

Although the population in level 1 gives the appearance of sinusoidal Rabi oscillations, Yeh, et. al. [12] have shown that the apparent sinusoid actually consists of a piecewise continuous sequence of exponential decays--or linear segments on a semilogarithmic plot of population versus time. The initial decay occurs during the first recurrence time interval $0 < t < \tau$, when the time bandwidth $2\pi/t$ greatly exceeds the detuning. This linear decay occurs according to the traditional Fermi Golden Rule at rate

$$R = 2\pi\rho V^2 \tag{4.7}$$

Here $\rho = 1/\delta$ is the density of energy levels and V^2 is, as in Eq. 4.4 the sum of the individual interaction squares. Using the definition of Eqs. 4.3 and 4.5 we can write

$$R = \pi^2\tau/t_R{}^2 = \pi\tau/(NT_R^2) \tag{4.8}$$

where, as above, t_R is the two-level Rabi period and T_R is the Rabi period appropriate to the full band of levels.

The population cannot continue to decay exponentially at rate R indefinitely, if for no other reason than the exhaustion of initial population. What actually occurs is rather curious. During the interval $0 < t < \tau$ the population decays at rate R; during interval $\tau < t < 2\tau$ it decays at rate 3R, etc. [12]. Figure 11

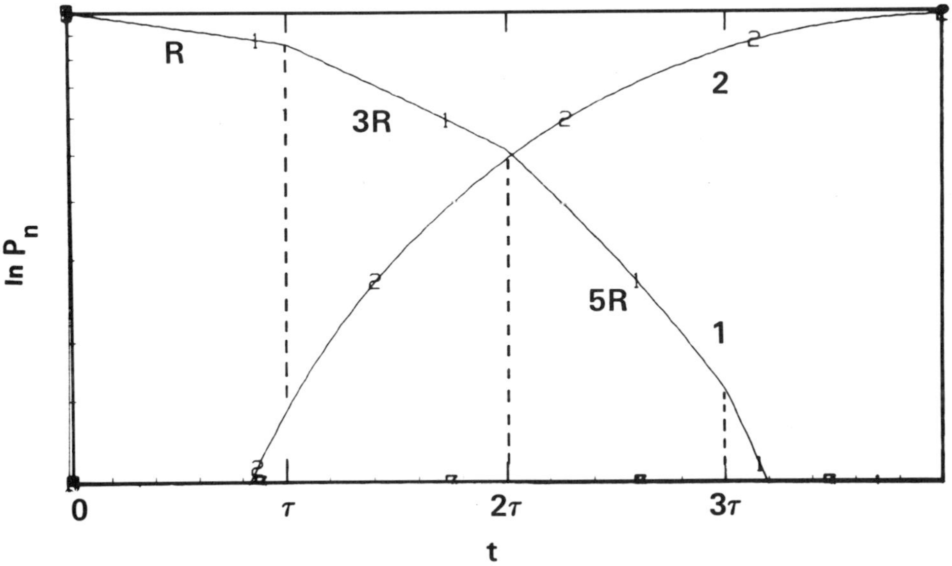

Figure 11
Plot of $\log P_n$ versus t, for parameters $\tau = 5$, $t_R = 40$ (and $T_R = 8$) showing breaks in exponential decay at $t = \tau, 2\tau, 3\tau$.

illustrates this behavior. We see here that the piecewise exponential behavior ceases for time exceeding half the two-particle Rabi period, $t > t_R/2$. This terminus applies so long as

$$\tau < t_R/2 .$$

The exponential decay can be prolonged to longer times up to one Rabi period $t = t_R$, by increasing τ to the value t_R.

As τ ranges over values from $\tau \ll t_R$ to $\tau \gg t_R$ the population histories P_n exhibit a variety of behavior, as shown in Fig. 9. When $\tau \ll t_R$, as is the case

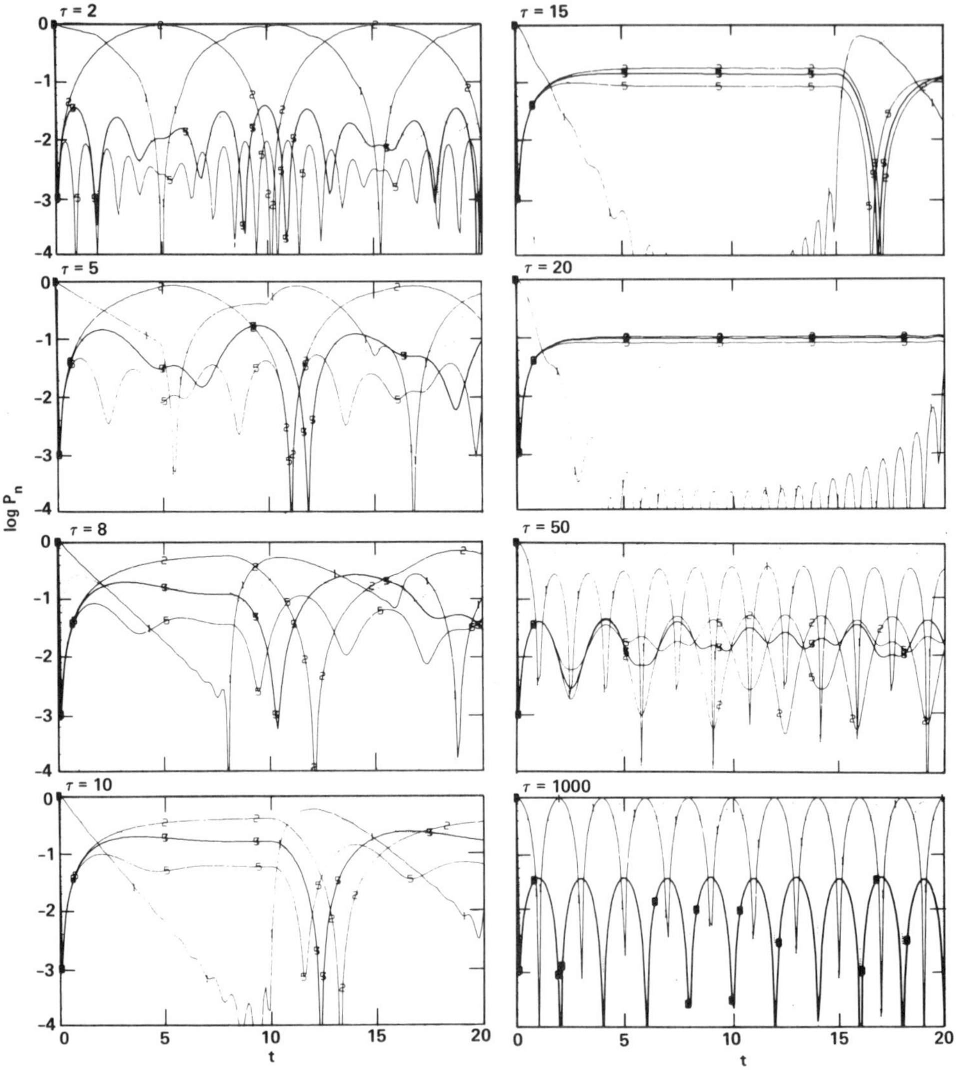

Figure 12
Plots of log P_n versus t for N = 25 level quasicontinuum having parameters t_R = 10 (and T_R = 2) and various values of repetition period.

for the first frame where τ = 2 and t_R = 10, the history of level 1 consists of a very large number of exponential decays, and the curve becomes indistinguishable from a sinusoid. When τ and t_R are approximately equal the exponential decay persists until the time t = t_R, after which time the population undergoes near periodic variation. When the recurrence period τ exceeds t_R then the exponential decay curve becomes sinusoidally modulated. Finally, when τ greatly exceeds the band Rabi period T_R the behavior shows no sign of exponential decay. Thus for fixed laser power (i.e. fixed t_R) QC behavior is most pronounced for level spacings such that $\tau \cong t_R$ (or $\delta \cong v/2$) and for times $t < \tau$

The long time average probabilities $\bar{P}_n$ also exhibit regularities as one varies τ. For small τ the populations of levels 1 and 2 are equal and are each close to the value 0.5, while populations of other levels fall off sharply with increasing n. For large τ level 1 has average population 0.5, while the remaining N levels each have the population 0.5/N. For intermediate τ (such as $\tau = t_R$) level 2 is generally the most populous, with level 1 less populous and other populations falling gradually with n. An interesting situation occurs for $\tau = 0.6\ (t_R)^2/T_R$. With this choice for τ the populations approximately satisfy the equality

$$\bar{P}_1 \cong \bar{P}_2 \cong \bar{P}_3 \cong \cdots \bar{P}_n \tag{4.9}$$

out to roughly n = N-4.

In the preceding examples the laser frequency has been chosen to be resonant with the central level of the QC (level 2). This choice of laser frequency has little effect upon the excitation by intense radiation (such that $v \gg \delta$), but for weak fields ($v \ll \delta$) subsequent to one recurrence time the behavior is quite sensitive to laser detuning. For example, if the laser frequency falls midway between QC levels, as shown in Fig. 13, then during the second recurrence interval $\tau < t < 2\tau$ the ground level probability increases at rate R (rather than decrease at rate 3R as occurs for the resonant tuning of Fig. 11). Figure 14 illustrates, for an N = 7 level QC, the effect of laser tuning upon the dynamics. We see that the effect is most pronounced for recurrence times much shorter than the Rabi period t_R; in the degenerate limit of $t_R \ll \tau$ the detuning has no effect.

5. QUASICONTINUUM PROPAGATION

In the present section we extend the previous treatment of optically thin excitation to the treatment of pulse propagation. We show the appearance of Beer's Law attenuation and we exhibit pulse echos. The same two elementary parameters characterize our study of propagation: the two-level Rabi frequency 2v (i.e. the dipole coupling between level 1 and any one of the QC levels) and the QC level spacing δ. These parameters enter most naturally when regarded as a (two-level) Rabi period $t_R = \pi/v$ and a recurrence time $\tau = 2\pi/\delta$.

We have already mentioned that QC behavior, as manifested by exponential decay of ground-level population, occurs most prominently when these two times are roughly comparable. To exhibit QC behavior most distinctly, we chose values $\tau = 1$ and $t_R = 2$, so that one Rabi period encompassed two recurrence times. Figure 15A shows the population behavior for this choice of parameters, under the assumption of monochromatic excitation starting at t = 0. This figure shows quite clearly the exponential decay of level 1 up to time $t = 1 = \tau$. As discussed in preceding sections, the decay rate for conditions of monochromatic excitation is, according to Fermi's Golden Rule, given by Eq. 4.8 and the population in level 1 follows approximately the decay law

$$P_1(t) = P_1(0)\exp(-Rt) = \exp(-Rt) \tag{5.1}$$

during the interval $0 < t < 1$.

To examine propagation we assumed a short pulse, of duration much briefer than the repetition time τ, so that during the excitation time interval the conditions produce exponential population decay. Figure 15B shows the assumed electric field envelope: a Gauss-10 pulse[12] of FWHM 0.2 and area $\pi/8$. Figure 15C shows the resulting population variations subject to this pulse.

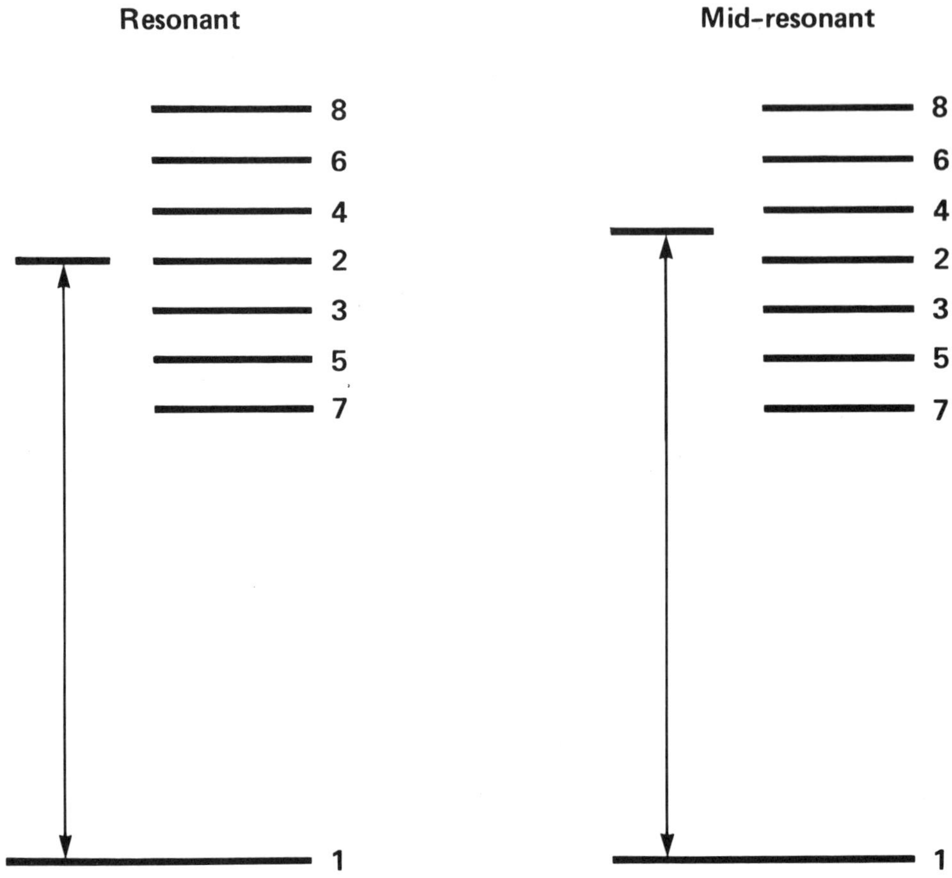

Figure 13
Laser tuning for resonant (left) and midresonant (right) QC excitation.

As this figure shows, once the pulse has passed the molecules at the entry face of the vapor, they remain in a distribution of excited states and hence give rise to a superposition of oscillating dipole moments. In turn these moments act as sources for an electromagnetic field. Each of the QC dipole moments has a different phase, and a frequency fixed by the detuning. At $t = 0$ these moments are in phase, but after a short time these different phases cancel, so that the total molecular dipole moment vanishes. After one recurrence time these dipoles will be in phase and hence can produce a total dipole moment. The result is a pulse echo. A second echo occurs after an additional recurrence time. Each such echo produces an impulsive diminution of the excited population by stimulating emission. Figure 15D shows these impulsive changes. They occur cyclically at multiples of the repetition rate τ.

Figure 16, showing the pulse intensity versus time t and propagation depth z, reveals two recurrence echos. The initial pulse fluence falls monotonically and in fact satisfies Beer's Law of exponential attenuation. To establish the Beer's

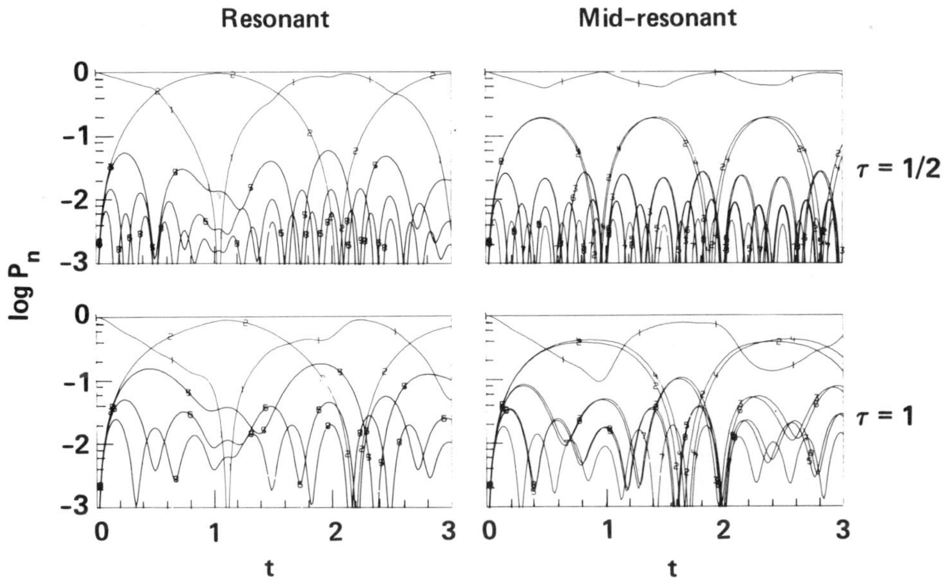

Figure 14
Population histories, as in Fig. 12 but for N = 7 level quasicontinuum. Left-hand frames are with resonant excitation (to level 2), right-hand frames are with mid-resonant excitation (between levels 2 and 4). Upper frames are for recurrence time $\tau = 1/2$, lower frames are for $\tau = 1$; all have Rabi period $t_R = 2$.

Law property we can examine plots of the unexcited population, $P_1(t,z)$ presented in Fig. 17. At any depth z the temporal behavior of $P_1(t,z)$ follows the pattern

$$P_1(t,z) = \exp\left[-\int_o^t R(t',z)dt'\right] . \qquad (5.2)$$

Here the rate R(t,z) is expressible in terms of pulse intensity I(t,z) as

$$R(t,z) = \sigma I(t,z)/h\nu . \qquad (5.3)$$

For a square pulse of duration τ_p we have the formula (for $t >> \tau_p$):

$$P_1(t,z) = \exp[-R(z)\tau_p] . \qquad (5.4)$$

If Beer's Law holds, so that the pulse fluence decays exponentially with optical depth $Z = \alpha z$ (where α is the Beer-Lambert attenuation coefficient), then

$$R(z) = R(0) \exp(-Z) . \qquad (5.5)$$

Combining these two expressions we obtain the expression

$$P_1(t,z) = \exp[-R(0)\tau_p \; e^{-Z}]$$

for the ground state population at any time after the passage of the pulse, $t > \tau_p$. It follows that

$$\ln[-\ln P_1(t,z)] = \ln[R(0) \; t_p]-Z . \qquad (5.7)$$

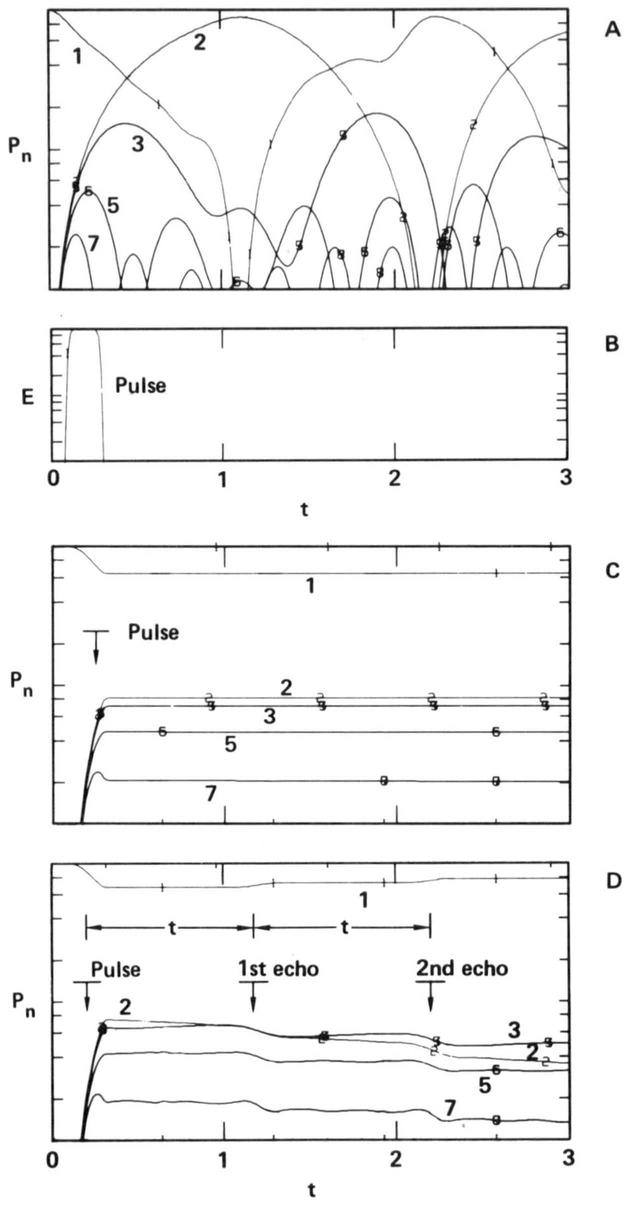

Figure 15
A) Population histories, as in Fig. 12, for monochromatic resonant excitation of N = 7 level quasicontinuum. B) Excitation electric-field pulse envelope. C) Population histories for excitation by pulse of B. D) Population histories after propagation.

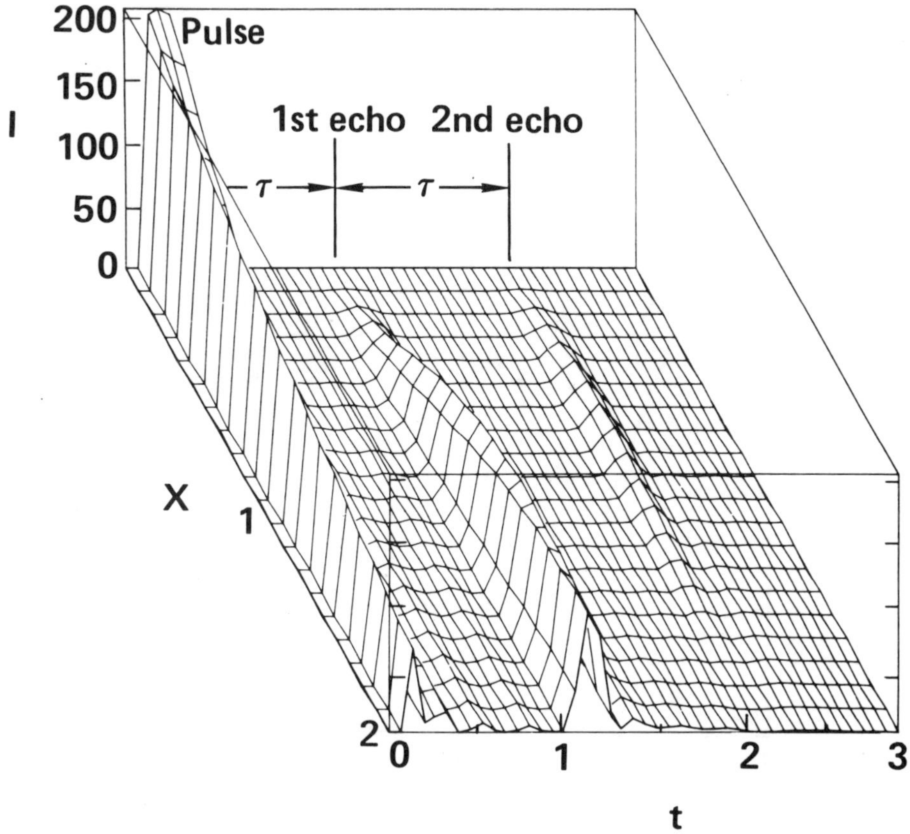

Figure 16
Pulse intensity versus time t versus Icsevgi-Lamb[14] propagation path parameter x for pulse of Fig. 15.

Thus a plot of $\ln(-\ln P_1)$ versus Z at any time after the passage of the pulse should yield a straight line of unit negative slope. Figure 17 shows such a plot for three selected values of t: at t = 1 the pulse has passed and the echo has yet to occur; at t = 2 the first echo has occurred; and at t = 3 the second echo has occurred. We see the predicted behavior.

It is interesting to observe the effect of lengthened repetition time τ upon the pulse echo. Figure 20 shows the echo, obtained with a resonantly tuned N = 7 level QC, for a range of recurrence times. In each case the Rabi period $t_R = 2$. For very short recurrence times (e.g., t = 1/4) the echo pattern consists of nearly equidistance pulses. As the recurrence time becomes longer than the pulse duration (e.g., $\tau = 1$) the echo pulses become well separated, and small secondary echoes become visible. These grow with increasing optical depth. As the recurrence time becomes longer than the Rabi period (e.g., $\tau = 4$), and the system consequently approaches 2-level degeneracy, the echo behavior becomes less significant, and we instead observe pulse reshaping. (In all these examples the initial pulse area is 0.2 π, so that soliton formation and self-induced transparency do not occur.)

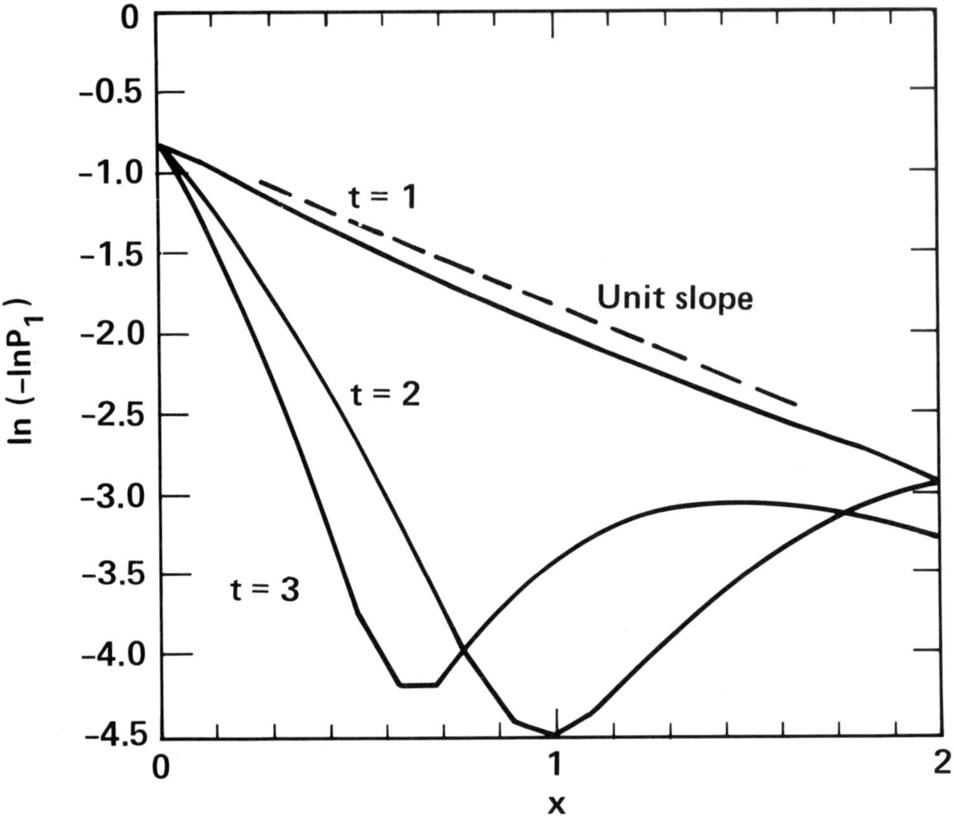

Figure 17
Plot of log (-log P_1) versus x for times t = 1, t = 2, and t = 3. Dashed line shows unit slope expected for Beer's Law attenuation.

The echo amplitudes are small compared with the initial pulse; we have therefore used a logarithmic plot to exhibit amplitudes.

It is also interesting to ask: how many levels are sufficient to reproduce the quasicontinuum echo behavior. To answer this question we exampled propagation with t_R = 2 and τ = 1 for various numbers of QC levels N. Figure 21 shows these results for resonantly tuned excitation (note that with N = 2 and N = 4 the levels are unsymmetrically distributed around the resonant level n = 2). We see that the echo is present as soon as there is a single "outrider" level added to the resonant level (the case N = 2). The addition of further outrider levels (N > 2) steadily sharpens the echo pattern and, in addition, introduces weaker shorter-recurrence echoes.

This behavior alters only slightly when we examine mid-resonant detuning. Furthermore, the primary echoes (those occurring at t = τ, 2τ, 3τ,...) are chiefly affected by the positions of the primary outrider levels (levels 3 and 4 when level 2 is resonant). The spacing of more distance QC levels has little effect upon either the width or the period of the echo pulses, so long as the

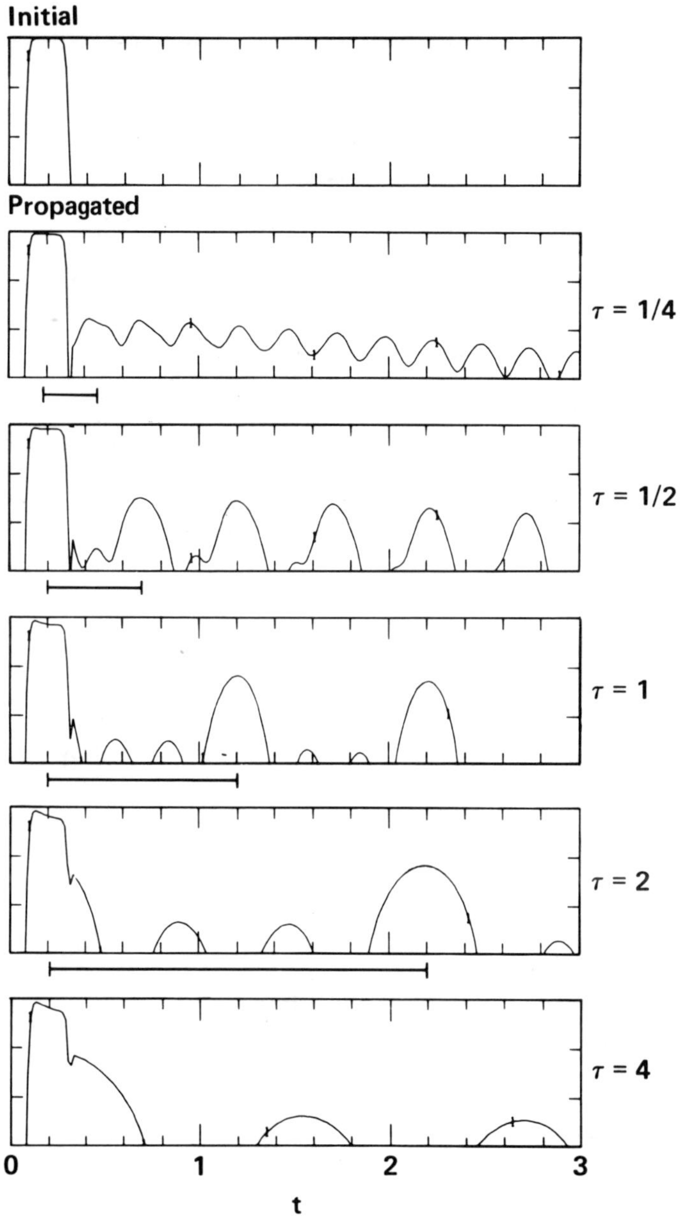

Figure 18
Pulse amplitude initially (top frame) and after propagation to path $x = 0.3$ (lower frames) for resonant excitation of $N = 7$ level QC with Rabi period $t_R = 2$ and various recurrence times τ.

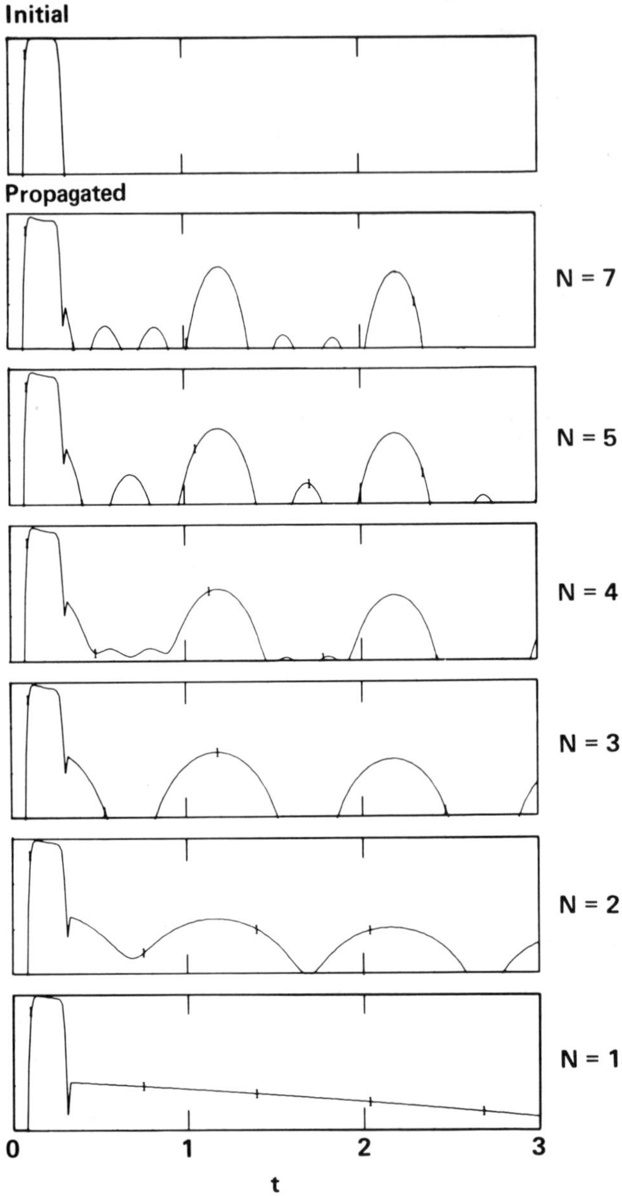

Figure 19
As in Fig. 18 with t_R and $\tau = 1$ and various numbers N of QC levels.

optical depth remains short enough that secondary echoes do not become dominant. We conclude that quasicontinuum behavior is evident for as few levels as N = 2.

6. SUMMARY AND CONCLUSIONS

The mutual coupling of radiation and matter provides an attractively realistic example of interacting nonlinear dynamical systems. It is important to realize that this is automatically true if the material system is near resonance with the radiation and is treated quantum mechanically. Classical treatments of this coupling (based on the Lorentz atom model) are intrinsically linear, unless the atomic polarizability is artificially endowed with nonlinear properties.

In this paper we have described the effects of spatial and temporal coupling of resonantly interacting radiation and matter. The three examples chosen were suggested by a desire to understand laser excitation of molecular vapor, a subject that has begun to be investigated intensively in the laboratory during the past five years.

It is difficult to find a mathematical model for a molecular vapor that is widely applicable in practice. We have reported here results for two commonly used models, and one new one. Our discussion of the new quasi-continuum model 12 is the first to include the reaction of the molecule back on the radiation; that is, the first to explore the effects of truly mutual nonlinear interaction, in that case. We have found evidence for behavior that we identify specifically with the quasi-continuum. This behavior is not well explained by either a purely discrete or a purely continuous model for the band of excited quantum states of the molecular vapor.

The most prominent example of quasi-continuum behavior appears in Fig. 16, where "echoes" of the input pulse follow each other at regular intervals. The degree to which the echoes reflect (new) nonlinear effects in the physics of radiation transport is shown in Fig. 17, where only the input radiation pulse (labelled t = 1) obeys the classical linear decay law associated with purely continuous models of the excited states. Of course a purely discrete model near to resonance would exhibit pronounced sinusoidal (Rabi) oscillations. However as Fig. 18 shows, the Rabi period is not nearly as important as the echo period, in describing the results of the interaction.

We have shown that numerical solutions of the coupled radiation-matter equations can illuminate such questions as those mentioned in the Introduction, regarding the presence of non-resonant levels and the number of them, the development of coherence, the essential features (intrinsic scales for time and distance) of propagation, etc. However, further work is required in several directions before general conclusions can safely be drawn. To mention only one example, it will be necessary to work with much more general quasi-continuum models. We have here assumed a strictly uniform spacing of energy levels in the band of excited states. Work is in progress on more general spacings and will be reported in a separate paper.

REFERENCES

1. In classical mechanics a purely harmonic oscillator responds purely linearly to an external force. Only those quantum mechanical systems with exactly equally spaced energy levels, infinite in number, have this property. We will discuss only models of molecules characterized by a finite number of energy levels. Our entire discussion thus necessarily deals with nonlinear molecular oscillators.

2. See, for example, L. Allen and J. H. Eberly, Optical Resonance and Two-Level Atoms (Wiley, New York, 1975), Chap. 4.

3. E. L. Hahn and S. L. McCall, Phys. Rev. 183, (1969) 457; G. L. Lamb, Jr., Elements of Soliton Theory (Wiley, New York, 1980), Chap. 7; and Rev. Mod. Phys. 43, (1971) 99.

4. A. A. Makarov, C. D. Cantrell, and W. H. Louisell, Optics Comm. 31, (1979).

5. B. W. Shore and J. Ackerhalt, Phys. Rev. A 15, (1977) 1640; J. H. Eberly, B. W. Shore, Z. Bialynicka-Birula, and I. Bialynicki-Birula, Phys. Rev. A 16, (1977) 2038; Z. Bialynicka-Birula, I. Bialynicki-Birula, J. H. Eberly, and B. W. Shore, Phys. Rev. A 16, (1977) 2048; B. W. Shore and J. H. Eberly, Optics Commun. 24, (1978) 83; R. J. Cook and B. W. Shore, Phys. Rev. A 20 (1979) 539; B. W. Shore and R. J. Cook, Phys. Rev. A 20 (1979) 1958.

6. J. H. Eberly, M. J. Konopnicki, and B. W. Shore, Optics Comm. 35 (1980) 76.

7. C. D. Cantrell, F. Rebentrost, and W. H. Louisell, Optics Comm. 36 (1981) 303.

8. B. Segard and B. Macke, Optics Comm. 38 (1981) 96; and B. Macke and F. Rohart, Optica Acta (in press).

9. G. B. Berman and G. W. Zaslavsky, Physica 2D (1981) 25.

10. M. Quack, J. Chem. Phys. 69 (1978) 1282; A. A. Makarov, V. T. Platonenko, and V. V. Tyakht, Sov. Phys. J.E.T.P. 48 (1978) 1044; C. D. Cantrell, V. S. Letokhov, and A. A. Makarov, in Coherent Nonlinear Optics, edited by M. S. Feld and V. S. Letokhov (Springer, Heidelberg, 1980), Chap. 5.

11. See, for example, Ref. 2, Chap. 3.

12. J. J. Yeh, C. M. Bowden, and J. H. Eberly, Phys. Rev. A (to be submitted).

13. By a "Gauss-10" pulse we mean one whose envelope at $z = 0$ is proportional to $\exp\{-[t-t_o)/w]^{10}\}$. Such a pulse is, in effect, both adiabatic (with smooth rise and fall) and square.

14. A. Icsevgi and W. E. Lamb, Jr., Phys. Rev. 185 (1979) 517.

COUPLED NONLINEAR OSCILLATORS
J. Chandra and A.C. Scott (eds.)

STATISTICAL ANALYSIS OF LONG-TERM DYNAMIC IRREGULARITY IN AN EXACTLY SOLUBLE QUANTUM MECHANICAL MODEL

F. T. Hioe, H.-I. Yoo and J. H. Eberly

Department of Physics and Astronomy
University of Rochester
Rochester, New York 14627
U.S.A.

We have studied the dynamics of the Jaynes-Cummings model after a long time from a statistical point of view. Even though the dynamical record of the Jaynes-Cummings variables such as atomic inversion become highly irregular in appearance, our analysis has revealed that there exists a mean angular frequency which characterizes the oscillations in both cases when the initial photon distributions are coherent and chaotic. We have also determined the partial recurrence frequencies of the atomic inversion. When the oscillation time of the dynamical function is scaled according to this mean angular frequency, and the amplitudes of oscillations scaled by their root mean square value, the behavior of the dynamical function is shown to be remarkably independent of the mean photon number and the initial photon distribution.

1. INTRODUCTION

The dynamics of a two-level atom (or spin one-half fermion) interacting, close to resonance, with a radiation field mode (or boson) has been of interest for many years.[1-7] The fundamental problem is dynamically nonlinear: In its mean-field form it is the coupled Maxwell-Bloch problem.[7] In its fully quantum mechanical aspect, i.e. with the radiation field quantized, the problem is known as the Jaynes-Cummings problem. Jaynes and Cummings found the stationary states and exact eigenvalues[1] of the coupled atom-field system twenty-five years ago. Very few studies of its dynamics were made[3,4] until the recent discovery by Eberly, Narozhny and Sanchez-Mondragon[5] that the atomic variables show unexpected periodic "collapses" and "revivals".

The most important features of the Jaynes-Cummings model are that it is fully quantum mechanical and exactly soluble. The solutions are therefore not restricted to domains of small coupling constant or weak excitation and are not subject to approximate decorrelations or truncations of a mean-field type. However, it has been important historically that the Jaynes-Cummings model has a well-studied mean-field analog (in which the field is not quantized) which is also free of restrictions on coupling constant, for which analytic solutions are known.

More than fifteen years after the work of Jaynes and Cummings, Ackerhalt[4] opened a new range of dynamical studies with the discovery that the fully quantum mechanical, apparently nonlinear, Heisenberg equations of motion of the model can be replaced by two coupled linear differential equations for the dipole moment and electric field operators. Ackerhalt was able to solve these equations and make some comparisons with the semiclassical solutions.

Most of the recent work has concentrated on the so-called "coherent-state" version of the model, in which the initial state of the radiation field is an eigenstate of the photon destruction operator.[2,3,5] This state is known to give a more realistic representation of experimental radiation fields than a pure photon number state. It has been found[5] that if the atom is in its ground state and the field is fully coherent initially, then the expectation value of the energy of the atom oscillates sinusoidally initially and then decays rapidly to a constant value, even though the model has no damping. Following the initial decay, or "collapse", there are periodic revivals to large amplitudes on a much longer time scale. On a still longer time scale, neighboring revivals begin to overlap, and the regions of overlap gradually include not only adjacent revivals but also more distant revivals, causing the envelope of the atomic inversion signal to appear increasingly irregular and noisy.[6] All of these features are shown in Fig. 1.

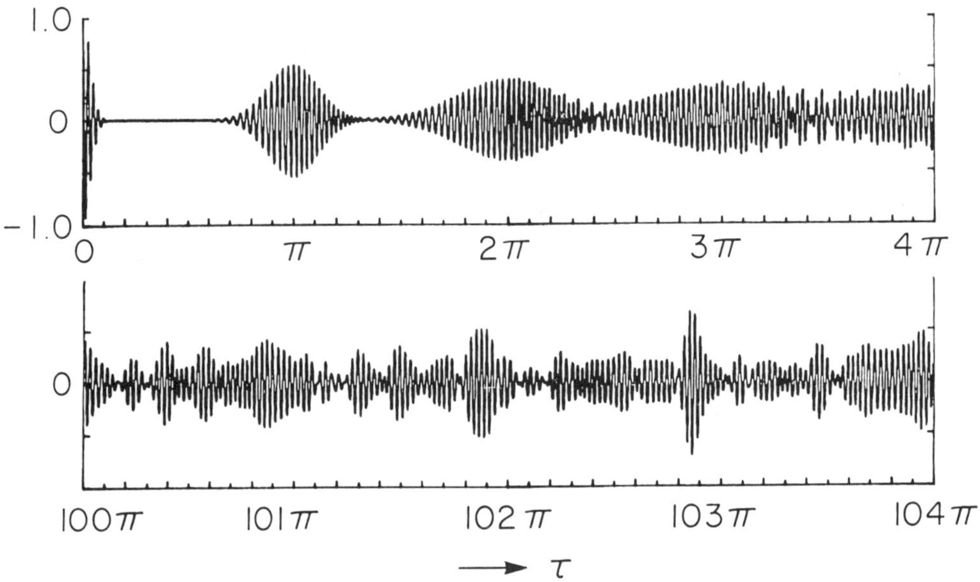

Figure 1
Time histories of the atomic inversion, $w(t)/m$ with the atomic spin down ($m = -1$) and the field in a coherent state with the average photon number $\bar{n} = 20$ as the initial state of the system. $\delta \equiv \Delta^2/4\lambda^2$ is chosen to be 0.01. The scaled time $\tau \equiv \lambda t/2\sqrt{\bar{n}}$, which was introduced in Ref. 4, has been used. The figure covers two widely separated time intervals of the same length: $0 \leq \tau \leq 4\pi$ and $100\pi \leq \tau \leq 104\pi$. The values $\tau = k\pi$ ($K = 1, 2, \ldots$) correspond to the peaks of the K-th revival signals.

The nature of this noisy irregularity, occurring at very long times, has not been previously studied. It is the main object of interest in this paper.

In this paper, we concern ourselves with properties of the dynamics of the Jaynes-Cummings model in the "irregular" regime. We ask what quantities should characterize such irregular behavior. Can the time dependent atomic response be approached from the statistical point of view? If so, is there a sensible "mean frequency" or oscillation even after a very long time? After a long time, does the atomic response to an initially coherent field become like the response to a field that was initially fully chaotic (thermal)? One of our results, as we show below, is that a statistical analysis reveals that a mean angular frequency can be defined for arbitrarily great times. Moreover, if the number of photons present is sufficiently large, the mean frequency coincides with the "Rabi frequency" normally used for the semiclassical version of the model.[1,7] Also, we find that the response to an initially coherent field is permanently different from the response to an initially chaotic field.

In our investigation we consider just these two cases:[8] (a) a "coherent" state where the initial photon distribution is Poissonian, and (b) a "chaotic" state with an initially thermal photon distribution. These two initial photon distributions constitute a complementary pair ("coherent" vs. "chaotic"). For large average photon number $\bar{n} \gg 1$, a Poissonian distribution is well localized around $\bar{n}$ with its width of order $\sqrt{\bar{n}}$, while a thermal distribution is broadly distributed with its width of order $\bar{n}$. Of course, at times after $t = 0$ both distributions change in accord with the exact atom-field interaction and are no longer either purely coherent or purely chaotic. We will continue to refer to them, however, as coherent or chaotic for simplicity.

The mathematical methods we use in this paper originated with the work of Lagrange, Wintner, Weyl and others in their pioneering study of perturbed planetary orbits.[9] Similar analyses have also been made of problems in molecular physics,[10] solid state physics and statistical mechanics.[11]

In Section II we review the basic features of the Jaynes-Cummings model problem. In Section III we introduce statistical concepts and define an appropriate probability density P(x) and evaluate its second moment exactly. An approximate expression for P(x) is derived in the limit of large photon number. Section IV is devoted to the mean angular frequency of the long-time irregular oscillation, and in Section V we consider the question of partial recurrence times for the model. Section VI is devoted to a summar, and an Appendix contains a number of mathematical details.

2. THE JAYNES-CUMMINGS MODEL AND THE ATOMIC INVERSION

The Hamiltonian representing an atom in a monochromatic electromagnetic radiation field of frequency ω, can be written as

$$\hat{H} = \hat{H}_F + \hat{H}_A + \hat{V} \tag{2.1}$$

where $\hat{H}_F$ is the Hamiltonian of the free electromagnetic field, $\hat{H}_A$ is the Hamiltonian of the atom. The energy of the interaction of the atom and the field is represented by $\hat{V}$, which can be written as

$$\hat{V} = -\hat{d}\cdot\hat{E}(\underline{r}_o) \tag{2.2}$$

where $\hat{d}$ is the atom's dipole moment operator and $\hat{E}(\underline{r}_o)$ is the electric field operator evaluated at $\underline{r}_o$.

In many problems connected with radiation, it is possible to consider only two energy levels of the system. The model of Jaynes and Cummings[1] describes an ideal system of one two-level atom (or a spin one-half fermion) with energy spacing $\hbar\omega_a$ and one quantized electromagnetic field mode (or a boson field) of frequency ω interacting with each other through the dipole interaction. The frequency ω of the electric field is assumed to be nearly coincident with the transition frequency connecting the atom's energy levels. If we label the two levels (+) and (-) and denote the corresponding eigenstates of H_A by $|+\rangle$ and $|-\rangle$ with energies $1/2\ \hbar\omega_a$ and $-1/2\ \hbar\omega_a$, the atomic operators, which operate in the two-dimensional Hilbert space spanned by these two state vectors, can be conveniently represented in terms of the Pauli matrices $\hat{\sigma}_1$, $\hat{\sigma}_2$, $\hat{\sigma}_3$ defined by

$$\hat{\sigma}_1 = \begin{pmatrix} 0 & 1 \\ 1 & 0 \end{pmatrix} , \hat{\sigma}_2 = \begin{pmatrix} 0 & -i \\ i & 0 \end{pmatrix} , \hat{\sigma}_3 = \begin{pmatrix} 1 & 0 \\ 0 & -1 \end{pmatrix} \tag{2.3}$$

or, in place of $\hat{\sigma}_1$ and $\hat{\sigma}_2$, $\hat{\sigma}_+$ and $\hat{\sigma}_-$ defined by

$$2\hat{\sigma}_\pm \equiv \hat{\sigma}_1 \pm i\hat{\sigma}_2 . \tag{2.4}$$

The electromagnetic field is represented as usual by the photon (boson) creation and annihilation operators $\hat{a}^+$ and[7] $\hat{a}$. In terms of these operators, the Hamiltonian of the Jaynes-Cummings model is[7]

$$\hat{H} = \hbar\omega\hat{a}^\dagger\hat{a} + \frac{1}{2}\hbar\omega_a\,\hat{\sigma}_3 + \hbar\lambda(\hat{\sigma}_+\hat{a}+\hat{a}^\dagger\hat{\sigma}_-) \tag{2.5}$$

where the operators satisfy the usual commutation relations for equal time:

$$[\hat{\sigma}_3,\hat{\sigma}_\pm] = 2\hat{\sigma}_\pm , \ [\hat{\sigma}_+,\hat{\sigma}_-] = \hat{\sigma}_3 , \ [\hat{a},\hat{a}^\dagger] = 1 , \tag{2.6}$$

$$[\hat{a},\hat{\sigma}_i] = 0 , \ [\hat{a}^\dagger,\hat{\sigma}_i] = 0 . \tag{2.7}$$

Eq. (2.1) represents the Hamiltonian of a model system consisting of just one two-level atom[7] in a quantized radiation field. The so-called rotating wave approximation[7] has been invoked. The so-called counter-rotating terms, $\hbar\lambda(\hat{\sigma}_+\hat{a}^+ + \hat{a}\hat{\sigma}_-)$, which also are implied by the fundamental interaction $-\vec{d}\cdot\vec{E}$, but which make no near resonant contribution, have been neglected. The coupling constant λ is taken to be real and positive. $\hat{H}$ can be decomposed into two parts which commute with each other, $\hat{H} = \hat{H}_1 + \hat{H}_2$:

$$\hat{H}_1 = \hbar\omega(\hat{a}^\dagger\hat{a} + \frac{1}{2}\hat{\sigma}_3) \tag{2.8a}$$

and

$$\hat{H}_2 = \frac{1}{2}\hbar\Delta\hat{\sigma}_3 + \hbar\lambda(\hat{\sigma}_+\hat{a} + \hat{a}^\dagger\sigma_-) \tag{2.8b}$$

where $[\hat{H}_1,\hat{H}_2] = 0$, and the detuning Δ is defined by $\Delta = \omega_a - \omega$. This decomposition of[12] H is the key to all known exact solutions of the Jaynes-Cummings dynamics.

We choose, at t = 0, a separable form for $\hat{\rho}(0)$:

$$\hat{\rho}(0) \equiv \hat{\rho}_f(0) \otimes \hat{\rho}_a(0) , \tag{2.9}$$

in which $\hat{\rho}_f(0)$ and $\hat{\rho}_a(0)$ are the initial field and the initial atomic density matrices respectively. Further we assume that initially the atom is either in the upper state (m = +1) or in the lower state (m = -1), and we write the field part in terms of the Fock state representation:

$$\hat{\rho}_a(0) = |m\rangle\langle m| \tag{2.10}$$

and

$$\rho_f(0) = \sum_k \sum_{k'} \langle\ |\hat{\rho}_f(0)|k'\rangle \cdot |k\rangle\langle k'| . \tag{2.11}$$

Here $|k\rangle$ is a Fock state with photon number k.

It is well known[12] that the exact solution for the atomic inversion $\omega(t) = \langle\hat{\sigma}_3(t)\rangle$, is given by:

$$\omega(t) = m_k \sum_{k=0}^{\infty} p_k \left\{ \frac{\Delta^2}{\Omega_k^2} + (1 - \frac{\Delta^2}{\Omega_k^2}) \cos \Omega_k t \right\} , \tag{2.12}$$

in which

$$p_k \equiv \langle k|\hat{\rho}_f(0)|k\rangle \tag{2.13}$$

is the diagonal element of the initial field density matrix and Ω_k is

$$\Omega_k \equiv 2\lambda\sqrt{k + 1 - \theta+\delta} \ . \tag{2.14}$$

In Eq. (2.14) we have defined

$$\delta \equiv \Delta^2/4\lambda^2 \tag{2.15}$$

and

$$\theta \equiv \frac{1}{2} (1-m) \ . \tag{2.16}$$

Obviously only the time-dependent part of $\omega(t)$ is dynamically interesting. We will designate this part by the term "atomic response function", and write it as:

$$x(t) \equiv \sum_{k=\theta}^{\infty} p_k \cos \Omega_k t \ . \tag{2.17}$$

The sum starts at $k = 0$ when the initial atomic spin is up ($m = +1$) and at $k = 1$ when it is down ($m = -1$). The temporal behavior of the atomic response function is obviously determined by the initial density matrix of the field through Eq. (2.13). The amplitudes p_k, in the two complementary cases of interest, are:

Coherent (Poisson)

$$p_k = \frac{(\bar{n})^k}{k!} e^{-\bar{n}} \tag{2.18}$$

Chaotic (thermal[13])

$$p_k = \frac{1}{1+\bar{n}} \left(\frac{\bar{n}}{1+\bar{n}}\right)^k \ . \tag{2.19}$$

3. THE PROBABILITY DENSITY P(x)

Even though X(t) is completely determined in (2.17) we can still discuss it statistically in the sense of relative frequency, as follows. A probability density, P(x), of the variable x can be introduced in a time averaged sense by the definitions:

$$\Delta M(x;\Delta x) \equiv \lim_{T\to\infty} \frac{1}{T} \Delta T(t:X\leq X(t)\leq X+\Delta X) \ , \tag{3.1a}$$

$$P(X) \equiv \lim_{\Delta X\to 0} \Delta M(X;\Delta X)/\Delta X) \ , \tag{3.1b}$$

where $\Delta T(t:X\leq X(t)\leq X+\Delta X)$ is the sum of time intervals in which X(t) stays between

X and X + ΔX for $0 \leq t \leq T$. P(x) may be computed as follows. The characteristic function $C(\alpha,X)$ of $\bar{X}(t)$, in the time-average sense, is:

$$C(\alpha;X) \equiv \lim_{T\to\infty} \frac{1}{T} \int_0^T \ell^{i\alpha X(t)}\, dt \,. \tag{3.2}$$

If the frequencies Ω_k are linearly independent, i.e., if no set of integers $\{m_k\}$ except $\{m_k=0\}$ can be found to satisfy

$$\sum_k m_k\Omega_k = 0 \,, \tag{3.3}$$

then we can use the Kronecker-Weyl (K-W) theorem[9] to replace the time average in (3.2) by the corresponding phase average. For this reason we assume δ is irrational and small from now on. This irrationality condition can be weakened if a small estimatable error to the final results is allowed. Then the K-W theorem gives:

$$C(\alpha;X) = \lim_{N\to\infty} \frac{1}{(2\pi)^{N+1-\theta}} \int_0^{2\pi} \cdots \int e^{i\alpha \sum_{k=\theta}^{N} p_k \cos\phi_k} d\phi_\theta \cdots d\phi_N \tag{3.4}$$

$$= \prod_{k=\theta}^{\infty} J_o(p_k\alpha) \,, \tag{3.4}$$

where $J_o(x)$ is the zero-th order Bessel function. The probability density P(x) of X(t) is the Fourier transform of the characteristic function $C(\alpha;X)$ and is therefore given by

$$P(x) = \frac{1}{2\pi} \int_{-\infty}^{\infty} \prod_{k=\theta}^{\infty} J_o(p_k\alpha) e^{-i\alpha x}\, d\alpha \,. \tag{3.5}$$

We show in the Appendix that, for n >> 1, P(x) is well-approximated by a Gaussian distribution:

$$P(x) = \frac{1}{\sqrt{2\pi\sigma}}\, e^{-\frac{1}{2}(x/\sigma)^2} \,, \tag{3.6}$$

where σ^2 is obviously the mean squared value of x. It can be computed easily from (2.17) by time averaging:

$$\sigma^2 = \frac{1}{2} \sum_{k=\theta} p_k^2 \,. \tag{3.7}$$

In the two cases that we consider in detail, σ^2 is given by:

Coherent

$$\sigma^2 = \frac{1}{2}\, e^{-2\bar{n}} \left[I_o(2\bar{n})-\theta\right] \cong (16\pi\bar{n})^{-1/2} \tag{3.8}$$

Chaotic

$$\sigma^2 = \frac{1}{2}\, \frac{1}{1+2\bar{n}} - \frac{\theta}{2}\, \frac{1}{(1+\bar{n})^2} \cong (4\bar{n})^{-1} \,, \tag{3.9}$$

where the approximate expressions are valid in the limit $\bar{n} >> 1$.

The limit $\bar{n} >> 1$ is adopted here for physical reasons. Only for large $\bar{n}$ does the coherent distribution lead to "coherent" dynamical behavior even for short times. That is, the atomic response X(t) exhibits a coherent sinusoidal time dependence (Rabi oscillations) for a time after t = 0 only if the distribution of p_k's is relatively sharply peaked at $k >> 1$. The coherent (Poisson) distribution has this property for $\bar{n} >> 1$. [The chaotic (thermal) distribution of course never has this property for any $\bar{n}$.] However, as we have shown elsewhere,[5,6] the coherent distribution eventually produces an irregular "chaotic" time record even for $\bar{n} >> 1$. Thus we concentrate here on the limit $\bar{n} >> 1$ in order to consider the question whether there are any significant remnants of an initially coherent atomic response X(t) that distinguish it, even after infinite time, from an X(t) that was chaotic from beginning.

One must also keep in mind the limits of the Gaussian approximation (3.6). The original atomic response function (2.17) is bounded by ±1, but (3.6) implies that X can be arbitrarily large. Equation (3.6) is realistic only if $|x| < 1$ and $\sigma << 1$, which is also consistent with $\bar{n} >> 1$, as is shown in (3.8) and (3.9). The accuracy of the Gaussian form is discussed in the Appendix.

The statistical basis for a study of atomic response is now established. In the next section we study the first long-time question for the model. That is, we determine the long-time mean frequency of response.

4. MEAN ANGULAR FREQUENCY

As is known from the pioneering work of Lagrange, Wintner, and Weyl, a central quantity in the study of quasi-periodic functions is the so-called "mean motion" or mean angular frequency. Let X(t) of Eq. (2.17) be the real part of a complex quantity Z(t) given by

$$Z(t) = \sum_{k=\theta}^{\infty} p_k e^{i\Omega_k t} \equiv r(t)e^{i\phi(t)} , \qquad (4.1)$$

where r and ϕ are the modulus and the phase angle of Z(t) in the complex plane.[14] Hence the quantity $\bar{\Omega}$ defined by

$$\bar{\Omega} \equiv \lim_{T\to\infty} \frac{1}{T}\{\phi(T)-\phi(0)\} , \qquad (4.2)$$

is the mean angular frequency with which the vector Z(t) rotates in the complex plane.

Weyl[9] showed that in case the Ω_k's are linearly independent, $\bar{\Omega}$ is given by the following weighted sum of Ω_k's:

$$\bar{\Omega} = \sum_k W_k \Omega_k \qquad (4.3)$$

with the use of Wintner's result:[14]

$$W_k = p_k \int_0^\infty J_1(p_k\alpha) \prod_{\ell\neq k} J_o(p_\ell\alpha)d\alpha . \qquad (4.4)$$

The integral in Eq. (4.4) can be evaluated approximately in the large $\bar{n}$ case by making the Gaussian approximation for the product of Bessel functions of zero-th order as we have done previously [Cf. (3.5) and (3.6)]. We find

$$W_k \cong p_k \int_0^\infty J_1(p_k\alpha)e^{-\frac{1}{4}\alpha^2 \sum_{\ell\neq k} p_k^2} d\alpha$$

$$\cong \frac{1}{2} p_k^2 \int_0^\infty e^{-\frac{1}{4}\alpha^2(2\sigma^2 - \frac{1}{2}p_k^2)} \alpha d\alpha \tag{4.5}$$

$$= \frac{p_k^2}{2\sigma^2 - \frac{1}{2}p_k^2}$$

and hence

$$\bar{\Omega} \cong \sum_k \frac{p_k^2 \Omega_k}{2\sigma^2 - \frac{1}{2}p_k} \tag{4.6}$$

or

$$\bar{\Omega} = \bar{\Omega}^{(0)} + \text{higher order terms} , \tag{4.7}$$

where $\bar{\Omega}^{(0)}$ is the leading contribution for $\bar{n} >> 1$. The large $\bar{n}$ limit can be evaluated, and we find:
Coherent

$$\bar{\Omega}^{(0)} = 2\lambda\sqrt{\bar{n}} \tag{4.8}$$

Chaotic

$$\bar{\Omega}^{(0)} = \frac{\pi}{2}\lambda\sqrt{\bar{n}} \tag{4.9}$$

There are two conclusions to be drawn from Eqs. (4.8) and (4.9). The first is that the leading contribution to the coherent mean angular frequency is precisely the semiclassical Rabi frequency. In other words, at times arbitrarily far from $t = 0$, the atomic response continues to oscillate with its mean frequency exactly equal to its initial frequency, in the "coherent" case, despite the apparent "incoherence" of the time record [see Fig. 1]. The second is that the atomic response to a chaotic initial field distribution has a mean angular frequency that is nearly the same as in the coherent case. It is perhaps the second conclusion that is more surprising.

In either case, Eqs. (4.8) and (4.9) suggest that there is a scaling with mean photon number $\bar{n}$ that can be used to define a "fundamental" time unit. This is shown in Fig. 2 for the coherent case, where $X(t)$ is plotted numerically, and time is measured in units of $1/\bar{\Omega}^{(0)}$. Both records, for very different values of $\bar{n}$, show clearly the same mean oscillation frequency.

What is more, the results of this Section can be combined with those of Sec. III to exhibit both amplitude and time scaling. It is clear from expression (3.6) for $P(x)$ that the expected amplitude of oscillation should scale with σ. That is, the statistical analysis predicts that the record of $X(t)$ should be invariant to changes in $\bar{n}$ if $X(t)/\sigma$ is plotted instead of $X(t)$, and if the time unit is $\bar{\Omega}^{(0)}t$ instead of t.

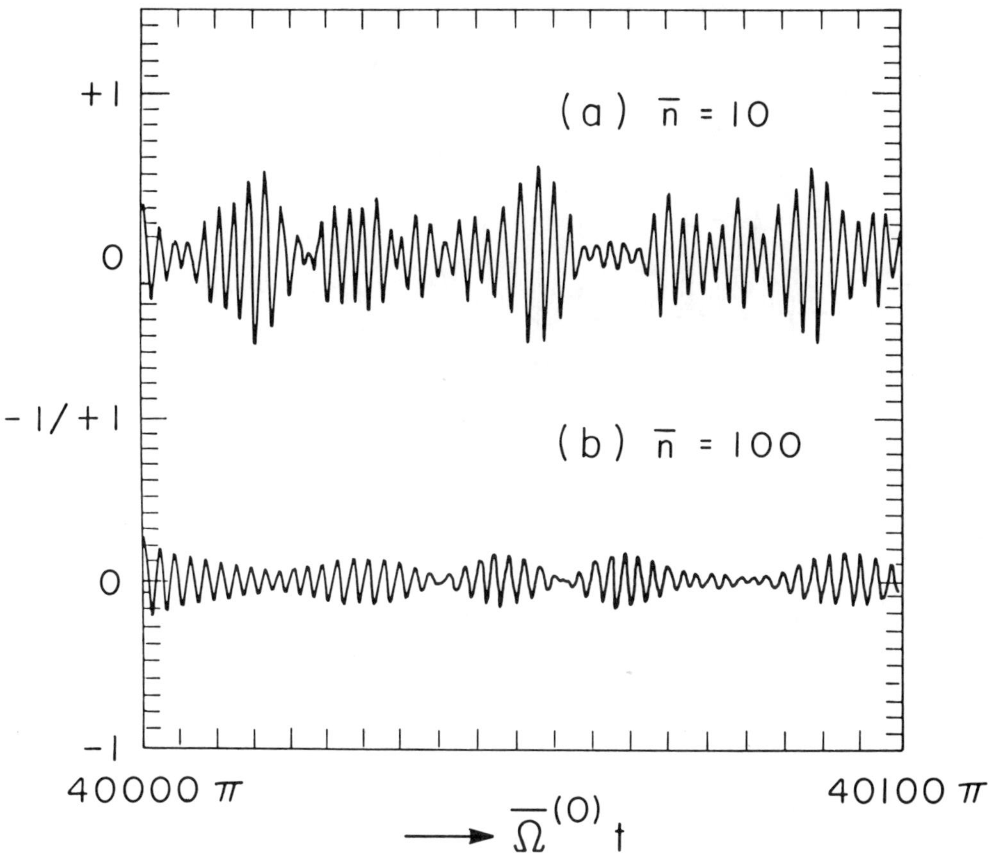

Figure 2
Very long time record of X(t) with Poissonian initial photon distributions p_k. Graphs (a) and (b) are for $\bar{n}$ = 10 and $\bar{n}$ = 100, respectively. The time interval shown in the figure is $40000\,\pi \leq \bar{\Omega}^{(0)}\,t \leq 40100\,\pi$ for both cases, and m = -1 and δ = 0.01, as in Fig. 1.

Figure 3 provides a test of this prediction. One sees in Fig. 3 the records of X(t)/σ vs. $\bar{\Omega}^{(0)}$t for three values of $\bar{n}$, for the chaotic case. The distributions of amplitudes are apparently very similar, in contrast to the distributions in Fig. 2, where X(t) is plotted on an absolute scale, not normalized by σ. An even stronger contrast is provided by Fig. 4, where the records of the same chaotic cases shown in Fig. 3 are replotted without either amplitude or time scaling.

5. PARTIAL RECURRENCES

The concept of mean angular frequency, discussed in the preceding section, may be loosely called a "local" concept. That is, if $\bar{\Omega}$ is a useful statistic it means that there is small dispersion in frequency, and $\bar{\Omega}$ is meaningful over even relatively short time intervals, not just over the infinite interval of its definition in Eq. (4.2). This concept of mean frequency can be substantially generalized, as follows.

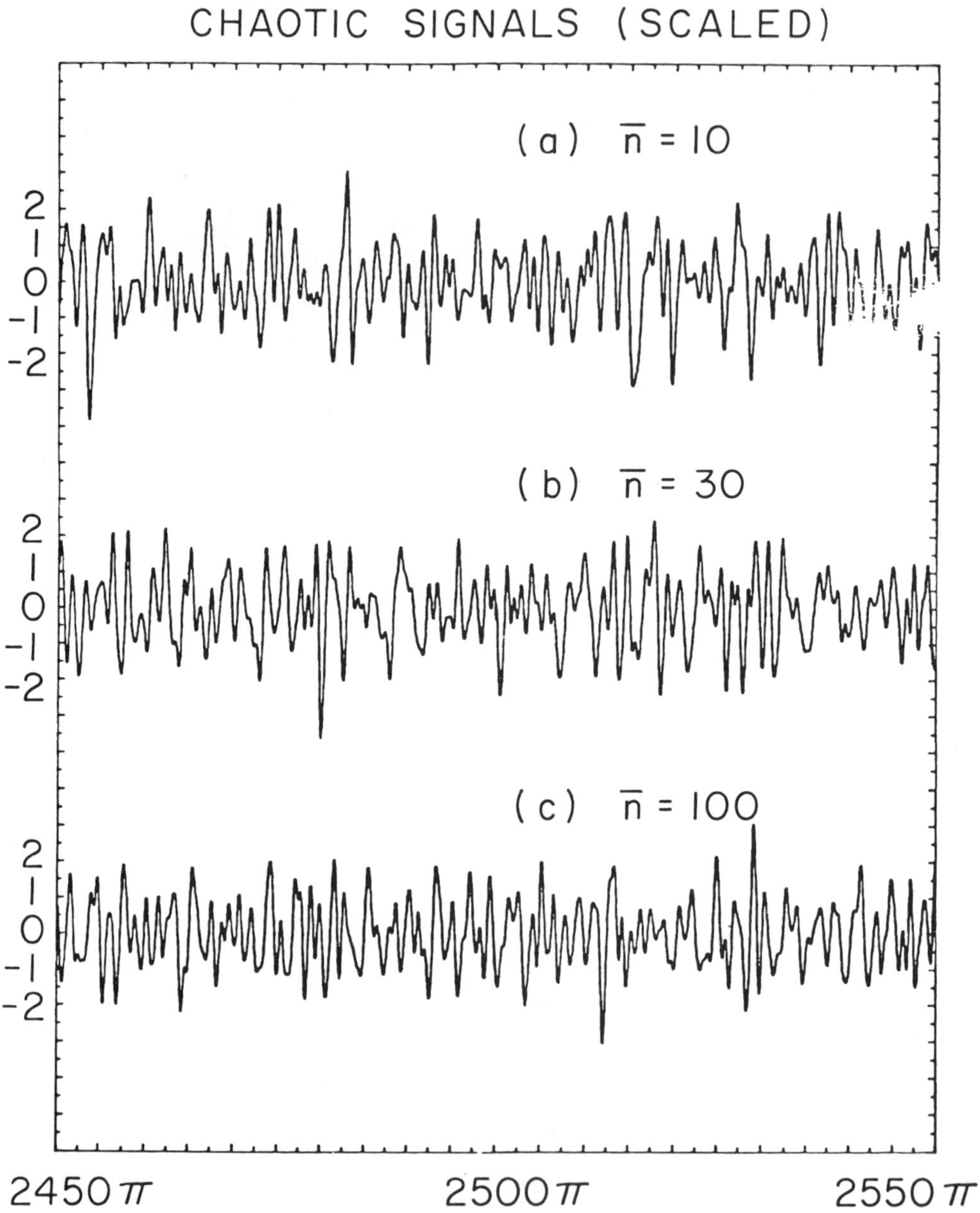

Figure 3
Portion of long-time record of $X(t)/\sigma$ with chaotic initial photon distributions p_k. Graphs (a), (b) and (c) are for initial average photon numbers $\bar{n}$ = 10, $\bar{n}$ = 30 and $\bar{n}$ = 100, respectively. The time interval shown in the figure is $2450\ \pi \le \bar{\Omega}^{(0)} t \le 2550\ \pi$ for all three cases. The values of σ and $\bar{\Omega}^{(0)}$ are $(\sigma, \bar{\Omega}^{(0)})$ = (0.14, 4.0 λ), (0.09, 6.9 λ) and (0.05, 12.5 λ) for the cases (a), (b) and (c), respectively, and m = -1, δ = 0.01, as before.

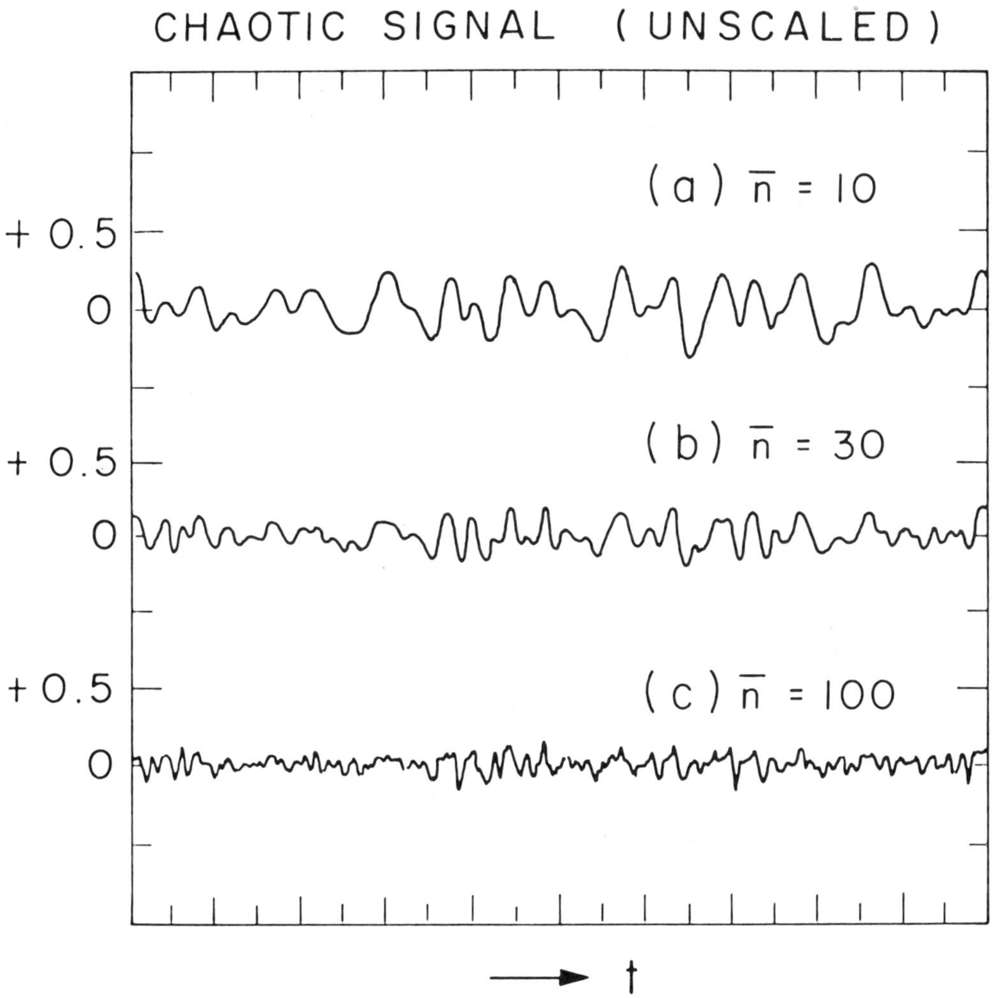

Figure 4
Portion of long-time record of X(t) with thermal p_k for (a) $\bar{n}$ = 10, (b) $\bar{n}$ = 30 and (c) $\bar{n}$ = 100. Both the vertical axis and the t axis are unscaled. The time range shown is $2450\ \pi/\bar{\Omega}^{(0)} \le t \le 2500\ \pi/\bar{\Omega}^{(0)}$ where $\bar{\Omega}^{(0)}$ is for $\bar{n}$ = 30; and m = -1 and δ = 0.01.

Let us consider how often the response function X(t) crosses the line x = q. If q is sufficiently different from zero, then the frequency of this event will be much smaller than the mean frequency $\bar{\Omega}$. These crossings will reflect very long-time features of the response function. We can speak of the event X(t) = 1 as a complete recurrence, or a return to the initial condition, and of the event X(t) = q < 1 as a partial recurrence. Of course X(t) will exhibit a complete recurrence only if it is periodic, which is not the case here.

One defines L(q), the average frequency of the partial recurrence X(t) = q, by

$$L(q) \equiv \lim_{T\to\infty} \frac{1}{T} N_T(q) \tag{5.1}$$

in which $N_T(q)$ is the total number of zeros of the function X(t) - q in the time range $0 \leq t \leq T$.

Kac[15] showed that when $X_N(t)$ is the sum of N sine waves of independent frequencies Ω_k, weighted with arbitrary positive coefficients p_k,

$$X_N(t) = \sum_{k=1}^{N} p_k \cos \Omega_k t \tag{5.2}$$

then L(q) is given by

$$L(q) = \frac{1}{2\pi^2} \int_{-\infty}^{\infty} \int_{-\infty}^{\infty} \frac{\cos q^{\alpha}}{\eta^2} \left[\sum_{k=1}^{N} J_o(p_k\alpha) - \underset{k=1}{\overset{N}{}} J_o(p_k\sqrt{\alpha^2+\eta^2\Omega_k^2}) \right] d\alpha d\eta \, . \tag{5.3}$$

We apply this formula for our case by taking the limit $N \to \infty$, and k starting at θ.

In the large $\bar{n}$ limit, the first product $\Pi J_o(p_k\alpha)$ on the right-hand side of Eq. (5.3) can be approximated by a Gaussian as we have shown in the Appendix. The second product can be approximated in the same way by appropriate scaling:

$$\Pi J_o(p_k\sqrt{\alpha^2+\eta^2\Omega_k^2}) \cong \exp\left(-\frac{\alpha^2}{4} \Sigma p_k^2\right) \exp\left(-\frac{\eta^2}{4} \Sigma p_k^2 \Omega_k^2\right). \tag{5.4}$$

The integration in (5.3) can then be performed and yields

$$L(q) \cong \frac{\bar{\Omega}}{\pi} \exp\left(-\frac{9^2}{2\sigma^2}\right) . \tag{5.5}$$

Here $\tilde{\Omega}$ is a new frequency unit defined by

$$\tilde{\Omega} = \frac{\tilde{\Omega}}{\sqrt{2\sigma}} \tag{5.6}$$

where

$$\tilde{\omega}^2 = \sum_{k=\theta}^{\infty} p_k^2 \Omega_k^2 \tag{5.7}$$

and where σ is given by Eq. (3.7). The sum can be evaluated, and one finds
<u>Coherent</u>

$$\tilde{\omega}^2 = 4\lambda^2 e^{-2\bar{n}} [\bar{n} I_1(2\bar{n}) + (1-\theta)I_o(2\bar{n}) + \delta \{I_o(2\bar{n})-\theta\}]$$

$$\cong 2\lambda^2\sqrt{\bar{n}}/\pi \, , \tag{5.8}$$

<u>Chaotic</u>

$$\tilde{\omega}^2 = 4\lambda^2\left[\left(\frac{\bar{n}}{2\bar{n}+1}\right)^2 + \frac{1-\theta}{2\bar{n}+1} + \delta\left\{\frac{1}{2\bar{n}+1} - \frac{\theta}{(\bar{n}+1)^2}\right\}\right]$$

$$\cong \lambda^2 , \tag{5.9}$$

where the leading term in the large - $\bar{n}$ limit is indicated in each case. Thus $\tilde{\Omega}$ in Eq. (5.5), for large $\bar{n}$, is given by

Coherent

$$\tilde{\Omega} \cong 2\lambda\sqrt{\bar{n}} \tag{5.10}$$

Chaotic

$$\tilde{\Omega} \cong \sqrt{2}\ \lambda\sqrt{\bar{n}} . \tag{5.11}$$

It is interesting to note that these expressions for $\tilde{\Omega}$ are of the same form as the semi-classical Rabi frequency. We notice that the quantity

$$\nu \equiv \tilde{\Omega}/2\pi = L(0)/2 \tag{5.12}$$

is the average frequency of crossing upward (or crossing downward) the $x = 0$ line by definition of $L(q)$. Hence $\tilde{\Omega}$ may be called the "average angular frequency of zero". Using this interpretation of $\tilde{\Omega}$, we see that a perfectly regular process would give $\tilde{\Omega} \equiv \bar{\Omega}$. Note that this relation does in fact hold true to first order for the "coherent" case [compare (4.8) and (5.10)], which is a strong indication of its actual continued coherence even in the face of strong long-term atom-field interactions. The failure of this relation for the chaotic case is not surprising that it is invalid by only 12%.

From Eq. (5.5) it is clear that the partial recurrence time $[L(q)]^{-1}$ grows larger than $\pi/\tilde{\Omega}$ very quickly for $q > \sigma$. It is of fundamental interest to consider $L^{-1}(q)$ for $q \cong 1$, i.e., to study the behavior near to full recurrence. Unfortunately, the Gaussian approximation is not accurate well before $q \cong 1$, and we have not found an alternative simple expression from Kac's formula (5.3) in the region $q \cong 1$. Nevertheless, even when applied to the limited record in Fig. 3, the prediction of (5.5) is verified within 8-10% for q in the neighborhood of 2σ.

6. SUMMARY

We have used well-known approximate methods of statistical physics, developed first to describe almost periodic planetary motion, to investigate the behavior of the Jaynes-Cummings quantum model. The model is one of the very few known quantum models that is exactly soluble for arbitrarily long times and arbitrarily large coupling constants.

The Jaynes-Cummings dynamics are deterministic and depend uniquely on whatever initial density matrix is chosen. There is no damping in the model. Nevertheless, after an interval of quasi-regular behavior, investigated in earlier papers, the dynamical record of a Jaynes-Cummings variable (such as atomic inversion or dipole moment, or radiation energy density) becomes highly irregular and random in appearance, as shown in Fig. 1.

We have computed both the mean motion (mean angular frequency) and the partial recurrence frequency of the inversion signal. For definiteness, and to obtain the largest contrasts, we have fixed our attention on two cases: the initial density matrix of the radiation was chosen to be completely coherent in one case and completely chaotic in the other. The evolution of the density matrix is of course dependent on its initial state.

Each case can be "scaled" by the mean square value of its response. These mean square values differ greatly (for large radiation densities) between the two cases. However, after scaling, the cases give remarkably similar results. This may be seen in (5.5), (5.10), and (5.11), for example, where the partial recurrence parameter $\tilde{\Omega}$ takes nearly the same value for coherent and chaotic initial conditions. This would be less remarkable if $\tilde{\Omega}$ reflected in some direct way the noisiness that is apparent in both cases at very long times. Instead $\tilde{\Omega}$ is precisely (in the coherent case) the so-called "semiclassical" Rabi frequency. That is, the very long time recurrence parameter is, in both cases, unexpectedly tied to the parameter that gives the frequency of the completely periodic oscillations of the approximate semi-classical solutions.

REFERENCES

1. E. T. Jaynes and F. W. Cummings, Proc. IEEE 51, 89 (1963).

2. F. W. Cummings, Phys. Rev. 140, A1051 (1965).

3. S. Stenholm, Physics Reports 6C, 62 (1973), and P. Meystre, E. Geneux, A. Quattropani, and A. Faist, N. Cim. 25B, 521 (1975), and references therein.

4. J. R. Ackerhalt, Ph.D. Thesis, University of Rochester (1974); and J. R. Ackerhalt and K. Rzazewski, Phys. Rev. A12, 2549 (1975), Sec. IV.

5. J. H. Eberly, N. B. Narozhny and J. J. Sanchez-Mondragon, Phys. Rev. Lett. 44, 1323 (1980); and Kvant. Elek. 7, 2178 (1980) [transl. in Sov. J. Quant. Electr. 10, 1261 (1981)]; N. B. Narozhny, J. J. Sanchez-Mondragon, and J. H. Eberly, Phys. Rev. A23, 236 (1981).

6. H. I. Yoo, J. J. Sanchez-Mondragon, and J. H. Eberly, J. Phys. A14, 1383 (1981).

7. See for example, L. Allen and J. H. Eberly, Optical Resonance and Two-Level Atoms (Wiley, New York, 1975), Ch. 7.

8. These two cases were discussed very early by F. W. Cummings in Ref. 2 without consideration of very long times.

9. H. Weyl, Math. Annalen. 77, 313 (1916), Am. J. Math. 60, 889 (1938), 61, 143 (1939) and references therein.

10. N. B. Slater, Proc. Cambridge Phil. Soc. 35, 56 (1936), N. B. Slater, Theory of Unimolecular Reactions (Cornell University Press, Ithaca, New York 1959), Ch. 4.

11. P. Mazur and E. W. Montroll, J. Math. Phys. 1, 70 (1960).

12. See, for example, Refs. 3 and 4.

13. The term "thermal" is associated with the chaotic distribution because the formula for p_k is the Bose Einstein formula, if $\bar{n}$ is interpreted as Planck's formula, for the average number of electromagnetic field oscillators that are excited at a given frequency and temperature.

14. A. Wintner, Am. J. Math. 55, 309 (1933).

15. M. Kac, Am. J. Math. 65, 609 (1943).

16. See, for example, I. S. Gradshteyn and I. W. Ryzhik, Table of Integrals, Series and Products (Academic Press, New York 1965), p. 14.

APPENDIX

Here we will give a systematic approximation scheme for the kernel of the integral in Eqs. (3.5) and (5.3) in terms of the average initial photon number $\bar{n}$, or, more precisely, the width of the initial photon distribution. The Gaussian forms Eqs. (3.6) and (5.5) are obtained as the lowest order approximaion in this scheme, but they are shown to be reasonably accurate for $\bar{n}$ as small as 5.

For the convenience of the analysis, we scale the variables p_k, α and X by σ given by Eq. (3.7);

$$p_k \to a_k \equiv p_k/\sigma$$

$$\alpha \to \beta \equiv \alpha\sigma \tag{A-1}$$

$$X \to y \equiv X/\sigma \; .$$

Equation (3.5) then becomes

$$P(x) = \frac{1}{2\pi\sigma} \int_{-\infty}^{\infty} K(\beta)\, e^{-i\beta y} d\beta \; , \tag{A-2}$$

where the kernel $K(\beta)$ is given by

$$K(\beta) \equiv \prod_{k=\theta}^{\infty} J_o(\alpha_k\beta) \; . \tag{A-3}$$

We notice that each Bessel function $J_o(\xi)$ exhibits oscillations with steadily decreasing amplitude as $|\xi|$ increases, starting at the maximum value of unity when $\xi = 0$. Hence the functional form of $K(\beta)$, which is a product of $J_o(\alpha_k\beta)$, has a localized peak around $\beta = 0$ with its maximum value of unity at the origin, and falls off rapidly as $|\beta|$ increases.

$K(\beta)$ remains positive until $\beta = \beta_c$ where the argument $\alpha_k\beta$ with the largest value of α_k reaches the first zero of $J_o^c(a_k\beta)$. In the region $0 \leq |\beta| < \beta_c$, we can use the following expansion:

$$K(\beta) = \exp\left[\sum_{k=\theta}^{\infty} \ln J_o(\alpha_k\beta)\right]$$

$$= \exp -\left[\sum_{\ell=1}^{\infty} A_\ell \beta^{2\ell}\right] \; . \tag{A-4}$$

The coefficients A_ℓ in Eq. (A-4) are given by[16]

$$A_\ell = B_\ell \sum_{k=\theta} \alpha_k^{2\ell} \tag{A-5}$$

with

$$B_\ell \equiv d_{\ell-1}/\ell\cdot 2^{2\ell} \; , \tag{A-6}$$

where $d_o = 1$

and for $n \geq 1$

$$d_n = (-)^n \begin{vmatrix} b_1 - c_1 & 1 & 0 & \cdots & 0 & 0 \\ b_2 - c_2 & b_1 & 1 & \cdots & 0 & 0 \\ b_3 - c_3 & b_2 & b_1 & \cdots & 0 & 0 \\ \cdots & \cdots & \cdots & \cdots & \cdots & \cdots \\ b_{n-1} - c_{n-1} & b_{n-2} & b_{n-3} & \cdots & b_1 & 1 \\ b_n - c_n & b_{n-1} & b_{n-2} & \cdots & b_2 & b_1 \end{vmatrix}$$

with

$$b_k = (-)^k/(k!)^2 \ , \quad c_k = b_k/(k+1) \ . \tag{A-8}$$

To obtain the formulas (A-5) - (A-8), we have used the fact that

$$\frac{d}{d\xi} \ell_n J_o(\xi) = - J_1(\xi)/J_o(\xi) \ , \tag{A-9}$$

and when the use the series expansions of J_o and J_1:

$$J_o(\xi) = \sum_{k=0} b_k(\xi/2)^{2k} \tag{A-10}$$

and

$$J_1(\xi) = \frac{\xi}{2} \sum_{k=0} C_k(\xi/2)^{2k} \ , \tag{A-11}$$

then the left-hand side quantity in Eq. (A-9) is expressed as

$$\frac{d}{d\xi} \ell n J_o(\xi) = - (\frac{\xi}{2}) \sum_{k=0} d_k (\frac{\xi}{2})^{2k} \ , \tag{A-12}$$

where b_k, c_k and d_k are given by Eqs. (A-7) and (A-8).

Since the determinant d_n, Eq. (A-7), consists of the sum of 2^n terms each of which does not exceed unity, we see immediately that the B_ℓ given by Eq. (A-6) are bounded as

$$|B_\ell| \leq 1/\ell \cdot 2^{\ell+1}$$

Actual values of B_ℓ for the first few ℓ's are

$$B_1 = 1/4 \ , \ B_2 = 1/64 \ , \ B_3 = 1/576 \ , \ B_4 = 11/3.2^{14}, \text{ etc.}$$

The width Δn of the initial photon distribution is $\Delta n \sim (\bar{n})^{1/2}$ when p_k is Poissonian, and is $\Delta n \sim \bar{n}$ when p_k is thermal. Equations (3.7) - (3.9) show that σ^2 is the same order as the inverse of the width $(\Delta n)^{-1}$ for both distributions for sufficiently large $\bar{n}$.

Moreover we can show that for sufficiently large $\bar{n}$ not only σ but also the sum $\sum_k a_k^{2\ell}$ in Eq. (A-5) can be characterized in terms of the powers of a parameter ε which is of the order of the inverse of the width $(\Delta n)^{-1}$, i.e.,

$$\sum_k a_k^{2\ell} \cong \frac{2}{\sqrt{\ell}} \varepsilon^{\ell-1} \tag{A-14}$$

with $\varepsilon \equiv 2/\sqrt{\pi\bar{n}}$

for Poissonian p_k, and

$$\sum_k a_k^{2\ell} \cong \frac{2}{\ell} \cdot \varepsilon^{\ell-1} \tag{A-15}$$

with $\varepsilon \equiv 4/\bar{n}$

for thermal p_k. In terms of this parameter ε, σ is $O(\varepsilon^{1/2})$. It is clear from Eqs. (A-5), (A-14) and (A-15) that the coefficient A_ℓ is of order $\varepsilon^{\ell-1}$. The first three values are

$$A_1 = 1/2\ ,\ A_2 \cong \varepsilon/32\sqrt{2} \text{ and } A_3 \cong \varepsilon^2/288\sqrt{3} \quad \text{for Poissonian } p_k\ , \tag{A-16}$$

and

$$A_1 = 1/2\ ,\ A_2 \cong \varepsilon/64 \quad \text{and} \quad A_3 \cong \varepsilon^2/864 \quad \text{for thermal } p_k\ . \tag{A-17}$$

Notice that A_1 is exactly 1/2 because of the scaling given by Eq. (A-1). Hence the coefficient of β^2 in the exponent in Eq. (A-4) is ε-free, while the remaining terms in the exponent become increasingly smaller as ε decreases (or as $\bar{n}$ increases). This feature is clearly seen in Fig. 5, where the first approximated form of $K(\beta)$, given by:

$$K_1(\beta) = e^{-A_1\beta^2}\ , \tag{A-18}$$

and the second approximated form of $K(\beta)$, given by:

$$K_2(\beta) = \exp(-A_1\beta^2 - A_2\beta^4)\ , \tag{A-19}$$

are directly compared with a numerical evaluation of the exact form of $K(\beta)$ given by Eq. (A-3) for three different values of $\bar{n}$, i.e., $\bar{n} = 1$, $\bar{n} = 5$ and $\bar{n} = 50$. The initial photon distributions p_k are Poissonian for the upper three figures [P 1], [P 2] and [P 3], and thermal for the lower three figures [T 1], [T 2] and [T 3]. In each of these figures Graph (A) represents $K(\beta)$ itself, (B) is $K(\beta) - K_1(\beta)$ and (C) $K(\beta) - K_2(\beta)$. We observe that both $K_1(\beta)$ and $K_2(\beta)$ approach $K(\beta)$ as $\bar{n}$ increases. The differences between $K_1(\beta)$ and $K_2(\beta)$ show the effect of the A_2 term, which is $O(\varepsilon)$, and becomes smaller as $\bar{n}$ increases.

In the figure we have ignored the restriction on the range of β and we have regarded Eqs. (A-18) and (A-19) [and (A-4) also] as the approximated forms of the true $K(\beta)$ for all values of β. The value of β_c is

$$\beta_c \cong 2.4/(a_k)_{max} \cong 2.4/\varepsilon^{1/2} \tag{A-20}$$

for both Poissonian and thermal distributions p_k when $\bar{n} \gg 1$.

Figure 6 [I] shows the comparison among $K_1(\beta)$, $K_2(\beta)$ and $K(\beta)$ for $\bar{n} = 30$. In Fig. 6 [II] the vertical axis is scaled up so that a more detailed comparison among several approximated kernels can be made, and Graph (D), showing $K(\beta) - K_3(\beta)$ is added, where $K_3(\beta)$ has a sixth order term in the exponent:

$$K_3(\beta) \equiv \exp(-A_1\beta^2 - A_2\beta^4 - A_3\beta^6)\ . \tag{A-21}$$

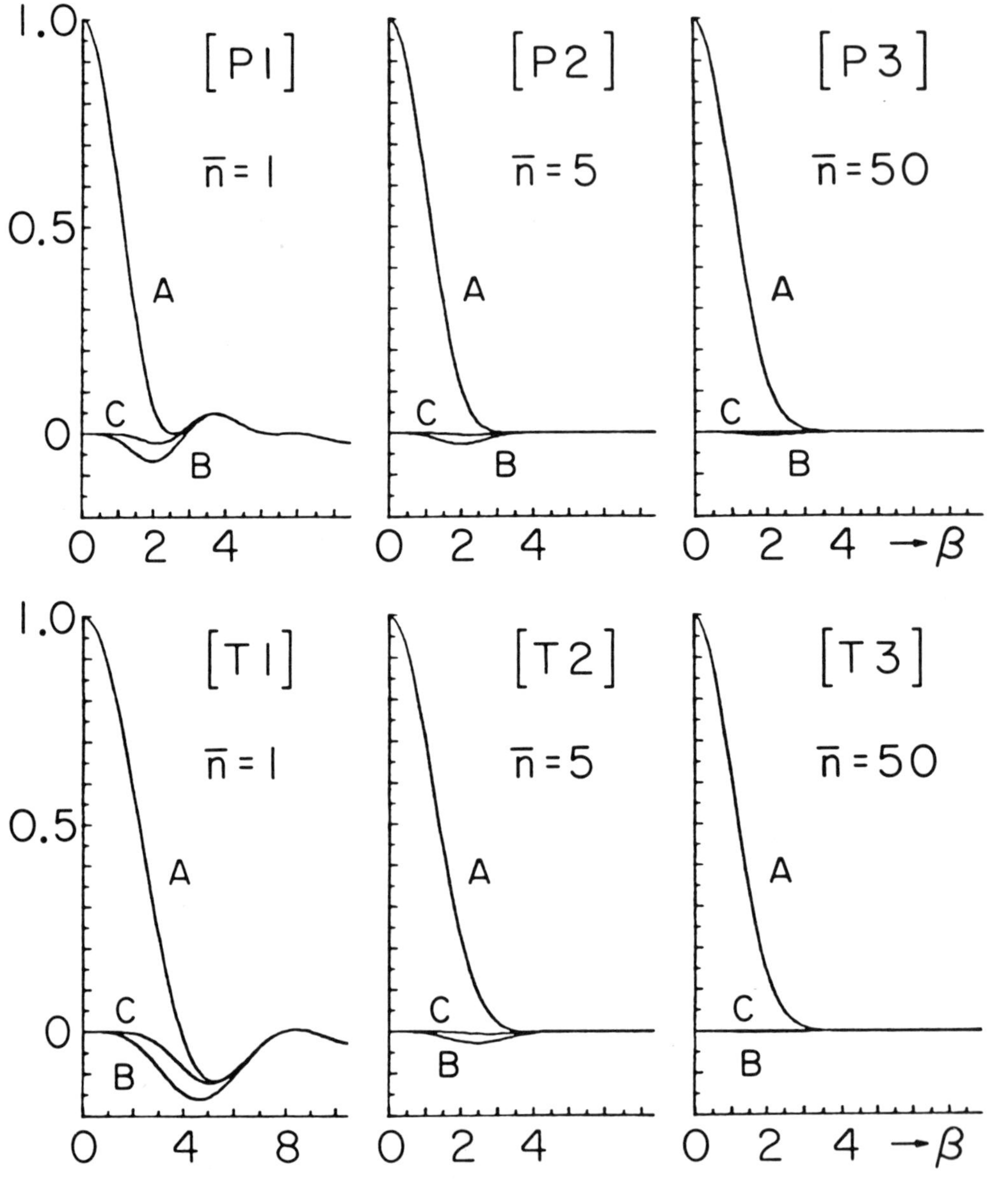

Figure 5

Comparison between $K(\beta)$ and its approximations $K_1(\beta)$ and $K_2(\beta)$ for three values of $\bar{n}$. Graphs (A), (B) and (C) are $K(\beta)$, $K(\beta) - K_1(\beta)$ and $K(\beta) - K_2(\beta)$, respectively. The upper three figures [P 1], [P 2], and [P 3] are for the Poissonian distributions p_k with $\bar{n} = 1$, $\bar{n} = 5$ and $\bar{n} = 50$, respectively. The corresponding figures for the thermal distributions p_k are shown in the lower three figures [T 1], [T 2] and [T 3]. $m = +1$ in all cases. The values of σ are 0.39, 0.25, 0.14, 0.20, 0.18, 0.07 for the cases [P 1], [P 2], [P 3], [T 1], [T 2], and [T 3], respectively.

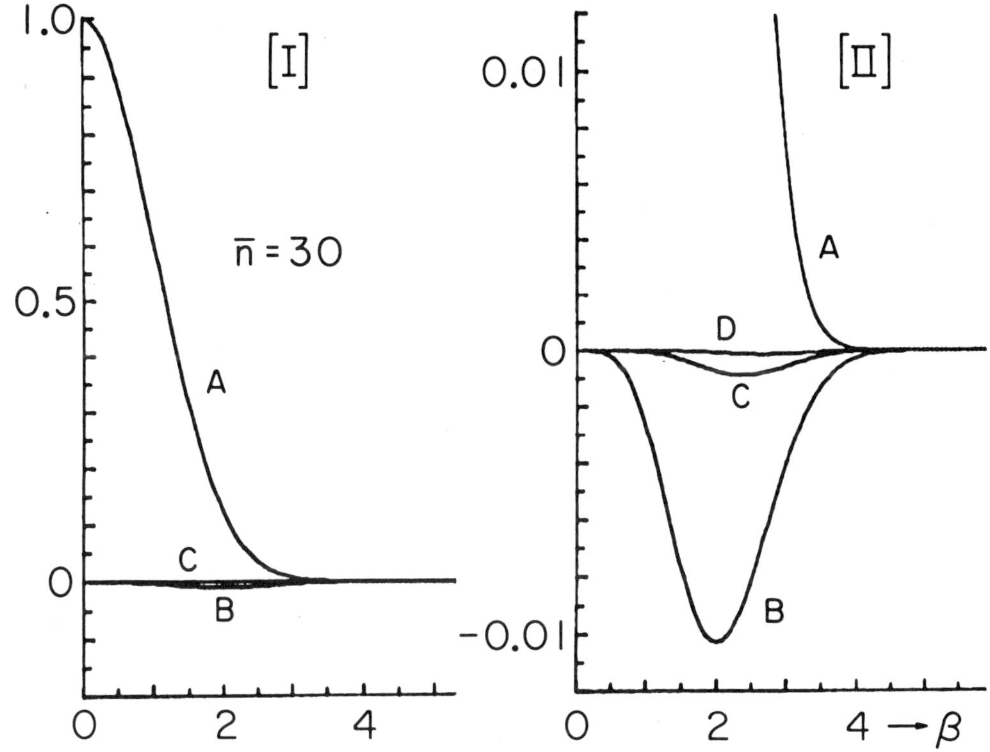

Figure 6
Comparison between $K(\beta)$ and approximated forms $K_1(\beta)$, $K_2(\beta)$ and $K_3(\beta)$. The initial photon distribution p_k is Poissonian with $\bar{n} = 30$. The left figure [I], shows (A): $K(\beta)$ (B): $K(\beta) - K_1(\beta)$ and (C): $K(\beta) - K_2(\beta)$. The right figure [II] shows a more detailed comparison, where (D): $K(\beta) - K_3(\beta)$. In all cases $m = +1$ $\sigma = 0.16$.

COUPLED NONLINEAR OSCILLATORS
J. Chandra and A.C. Scott (eds.)

ASPECTS OF INTERRUPTED COARSE-GRAINING IN STIMULATED EXCITATION OF VIBRONIC BANDS

C. M. Bowden

Research Directorate, US Army Missile Laboratory
US Army Missile Command
Redstone Arsenal, Alabama 35898

and

J. H. Eberly

Department of Physics and Astronomy
University of Rochester
Rochester, New York 14627

We consider the effect of interrupted coarse-graining at an arbitrary time after photo-excitation, and the dependence of certain dynamic discontinuities on the width of the vibronic band.

INTRODUCTION

Photo-excitation of a vibronic band of levels is a process of wide occurrence in experimental physics. Three examples are: (a) Broad-band flash pumping of upper laser levels in some solid and liquid laser media; (b) Narrow-band laser excitation of fluorescence in low-temperature glasses; and (c) Laser-induced photo-excitation of dissociative or pre-dissociative levels in polyatomic molecules. Although the theory that we discuss has some application to all of these examples, it is convenient to speak as if our attention were focused on a particular one of them. We adopt here the context of laser-induced predissociation. We hope that the reader who is more interested in another example can easily make the required translations.

In the first approximation, molecular bond vibrations are harmonic. That is, the restoring force that opposes molecular bond distortion is linearly proportional to the degree of distortion. Such an ideal molecular bond obviously never breaks. The molecule cannot be dissociated. Thus the problem of molecular dissociation is a problem of nonlinear mechanics, a problem of coupled nonlinear oscillators if the molecule is polyatomic. A quantum mechanical view of the dissociation problem gives a different perspective. The existence of nonlinearity and the possibility of dissociation both imply, and are implied by, the fact that the vibrational energy levels are equally spaced only for low energies and have an accumulation point at a certain energy, above which they are distributed continuously. This energy is the dissociation threshold.

An important element of a quantum mechanical view of dissociation is the obvious linearity of Schrödinger's equation. In a sense, a quantum mechanical approach allows a nonlinear problem to be traded for a linear one. Of course, it is possible to solve Schrödinger's equation for very few anharmonic potentials. However, this need not be a complete barrier to a quantum mechanical study of dissociation processes, depending upon which features of dissociation are deemed most interesting. A common view has been that it is interesting to study the dynamical process by which a given anharmonic molecular vibrational mode, during intense excitation by an external force (which we will identify with laser irradiation), becomes excited. One can associate this process, admittedly somewhat loosely, with the early stages of dissociation. In this paper, we assume that one discrete vibrational level is coupled to a closely-spaced band of higher levels. These levels serve in our model to simulate a real molecule's positive-energy final predissociation or dissociation levels.

It should be obvious that we have side-stepped the problem of solving the molecular structure equation for an anharmonic potential. That is, we have assumed that the positions and spacings of the anharmonic energy levels can be determined approximately in theory, or by appropriate spectroscopic experiments, and we take their existence for granted. We have not, however, made any assumptions about the molecular dynamics. The dynamical response of the molecule must be determined by solving the time-dependent Schrödinger equation.

We have previously considered[1,2] the photo-excitation of a band of discrete, but very closely spaced, vibrational energy levels. We have referred to the levels in the band as comprising a quasi-continuum[3] (QC). For simplicity, we have assumed that all of the levels in the QC are coupled via dipole matrix elements of roughly equal strength to a single lower-lying "ground" level, as shown in Fig. 1. The result of our earlier analysis[1,2] was the prediction of abrupt discontinuities in the molecule's dynamical response (see Fig. 2). These occurred at times that were identified as characteristic of the QC. Our analytic method was based on an ad-hoc sequence of coarse-grained integrations of the appropriate Schrödinger equation. In this paper we will investigate further features of the molecular response, and of our method of analysis.

MODEL OF LASER INDUCED QUASICONTINUUM EXCITATION

Figure 1 shows a model of quasicontinuum absorption. One photon is sufficient to reach the region of quasi-dense levels from the "ground" level $|0\rangle$. Of course $|0\rangle$ does not have to be the true ground level. Although each level $|m\rangle$ will have finite width due to normal line-broadening effects of several kinds (spontaneous decay, collisions, excitation into a higher QC, etc.) we assume the QC not so dense that these widths overlap each other, and we ignore such extrinsic relaxation phenomena.

We will take into account the entire QC band in estimating the rate of population loss from $|0\rangle$ during the excitation process. The Hamiltonian, H, which represents the system is

$$H = \hbar \sum_{r=0}^{N} \varepsilon_r R_{rr} - \underline{U}\cdot\underline{E} \ . \tag{1}$$

The first term on the right-hand side describes the molecular system with energy levels ε_r in the absence of laser field interaction. The expectation values of the operators R_{rr}, i.e., $\langle R_{rr}\rangle$, give the time dependent probability that the rth energy level is occupied, $0 \leq \langle R_{rr}\rangle \leq 1$. The second term on the right-hand side of (1) describes the interaction of the molecular system with the electric field of the laser in the electric dipole approximation, where the molecular dipole moment operator is given by

$$\underline{U} = \sum_{m=1}^{N} [\underline{d}_{mo} R_{mo} + \underline{d}_{om} R_{om}] \tag{2}$$

and the electric field $\underline{E}$, treated classically, is given by

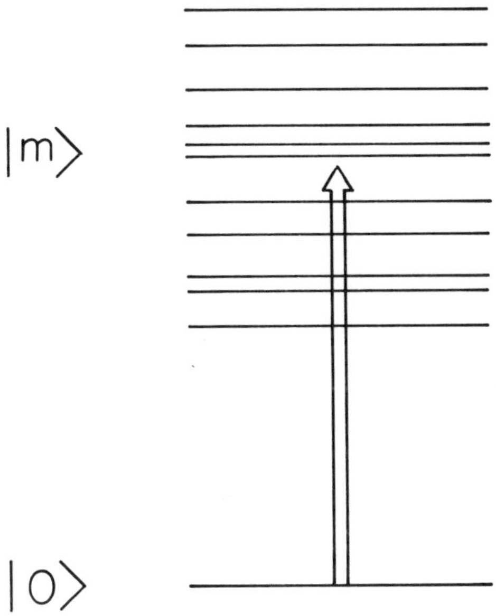

Figure 1
A ground-to-quasicontinuum transition is indicated schematically.

$$\underline{E} = \underline{E}^{(+)} + \underline{E}^{(-)} , \tag{3}$$

$$E^{+} = \hat{\underline{e}}\ \varepsilon(t)\ e^{-i\omega_L t} . \tag{4}$$

In the above relations, $\underline{d}_{mo} = \hat{\underline{e}}\ d_{mo}$ where d_{mo} is the matrix element of the transition dipole moment between the ground state |0> and the QC state |m> and $\hat{\underline{e}}$ is a unit vector taken to be along the field polarization direction, for convenience. The operators R_{mo}, R_{om}, are shift operators which connect states |0> and |m>, and $R_{mo} = R^{+}_{om}$. The complex electric field is represented by a slowly-varying envelope $\varepsilon(t)$ and carrier frequency ω_L.

In the rotating wave approximation,[4] the Hamiltonian becomes

$$H = \hbar \sum_{r=0}^{N} \varepsilon_r R_{rr} - \sum_{m=1}^{N} \underline{d}_{mo}\ \varepsilon(t)\ e^{-i\omega_L t} R_{mo}$$

$$+ \underline{d}_{om}\ \varepsilon^{*}(t)\ e^{i\omega_L t} R_{om} . \tag{5}$$

The operators $R_{k\ell}$ in (5) obey the Lie algebra[5]

$$[R_{ij}, R_{k\ell}] = R_{i\ell}\delta_{kj} - R_{kj}\ \delta_{i\ell} , \tag{6}$$

and all the dynamics of the system are determined from the Heisenberg operator equations of motion,

$$i\ \frac{\partial R_{kj}}{\partial t} = [R_{kj}, H] . \tag{7}$$

From (5) and (6), the set of $(N+1)^2$ equations generated by (7) are finite in number and linear. Thus, in principle, the set of dynamical equations of motion are exactly soluable, even for N large. Some simplication is achieved by noting from (6) that the molecular operators $R_{k\ell}$ have the realization

$$R_{k\ell} = a_k^+ a_\ell \tag{8}$$

where a_j are Bose operators

$$[a_k, a_j^+] = \delta_{kj} \ . \tag{9}$$

Thus, all the dynamics of the system can be obtained from the hierarchy of N+1 equations of motion,

$$i \frac{da_k}{dt} = [a_k, H] \ , \ k = 0,1,2,\ldots,N \ . \tag{10}$$

If (8) is used in (5), and the equations of motion calculated from (10), the result is

$$\dot{a}_o = -i \sum_{m=1}^{N} \beta_m^* e^{i\omega_L t} a_m \ , \tag{11a}$$

$$\dot{a}_m = -i \varepsilon_m a_m - i\beta_m e^{-i\omega_L t} \ , \tag{11b}$$

where $\cdot \equiv d/dt$ and we have set $\varepsilon_o = 0$. In Eqs. (11) the parameter β_m is half of the familiar two-level Rabi frequency.[4] That is

$$\beta_m = - \frac{\underline{d}_{mo} \cdot \hat{e}_L \ \varepsilon(t)}{\hbar} \ . \tag{12}$$

We note that Eqs. (11) are isomorphic to the equations of motion for the Schrödinger amplitudes. To see this, note that our Hamiltonian (5) with the realization (8) is bilinear in the Bose operators (9) and is of the general form

$$H = H_o + H_I$$

$$H_o = \hbar \sum_{m=0}^{N} \varepsilon_m a_m^+ a_m \ , \tag{13a}$$

$$H_I = \sum_{\substack{m,n=0 \\ m\neq 0}}^{N} g_{nm} \ a_n^+ a_m \ . \tag{13b}$$

Thus, using Eqs. (13) in (11), we find

$$i\hbar \frac{\partial a_n}{\partial t} = \varepsilon_n a_n + \sum_{\substack{m=0 \\ m\neq n}}^{N} g_{nm} a_m \ . \tag{14}$$

On the other hand, the equations of motion for the Schrödinger amplitudes $c_n(t)$ associated with the model's eigenlevels, are essentially the same

$$i\hbar \frac{\partial C_n(t)}{\partial t} = H_{nm} \ C_n(t) + \sum_{\substack{m=0 \\ m\neq n}} \langle n|H|m\rangle \ C_n(t) \ . \tag{15}$$

Equations (15) are for the Schrödinger amplitudes $C_n(t)$, whereas Eqs. (14) are the Heisenberg equations of motion for the operators $a_n(t)$. The correspondence is realized by taking the expectation value of the operator equations (14) with response to the initial state and the rule which defined the mapping is

$$\langle a_n(t)\rangle = C_n(t) \; . \tag{16}$$

From here on, we take the Eqs. (11) as c-number amplitude equations rather than operator equations.

It is convenient to invoke a simple transformation to a slowly-varying amplitude approximation defined by

$$a'_m = a_m e^{i\omega_L t} \quad m \neq 0$$

$$a'_o = a_o \; .$$

Then (11) becomes

$$\dot{a}_o = -i \; \Sigma \; \beta^*_m a_m \quad , \tag{17a}$$

$$\dot{a}_m = -i \; \Delta_m a_m - i\beta_m a_o \; , \tag{17b}$$

where we have dropped the prime on the variables in (17) for convenience, and have denoted individual transition frequency offsets from the laser frequency ω_L by Δ_m:

$$\Delta_m = \varepsilon_m - \omega_m \tag{18}$$

We will refer to the Δ's as de-tunings.

TIME-DEVELOPMENT KERNEL AND STATEMENT OF PROBLEM

We assume that, at $t = 0$, the molecule is in level $|0\rangle$. Thus, $a_m(0) = 0$ ($m \neq 0$ by convention throughout the paper), and we can easily solve (17b) formally. Without losing any dynamical features of present interest we can take the laser pulse to be very slowly-varying, and put $\beta_m(t) = \beta_m$ (constant). Then we obtain

$$a_m(t) = -i\beta_m \int_o^t dt' \; \exp[-i\Delta_m(t-t')]a_o(t') \; . \tag{19}$$

This permits (17a) to be rewritten:

$$\dot{a}_o = - \int_o^t dt' K(t-t')a_o(t') \; , \tag{20}$$

where the time-development kernel $K(t,t')$ is given by

$$K(t-t') = \sum_m \; [\exp - i\Delta_m(t-t')]\,|\beta_m|^2 \; . \tag{21}$$

In the remainder of this paper we will use the term quasicontinuum to mean that the time development kernel K has features that cannot be completely explained by assuming a continuous density of states of the excited band of levels labeled $|m\rangle$. Some features of K can only be explained by the discreteness of the levels m. We have identified both kinds of features in the course of our previous analyses, and an example is shown in Fig. 2.

We have shown that, for sufficiently short times, the dynamical response of the molecule is as if its levels were continuously distributed. This is not surprising, as the Heisenberg uncertainty principle does not permit sharp

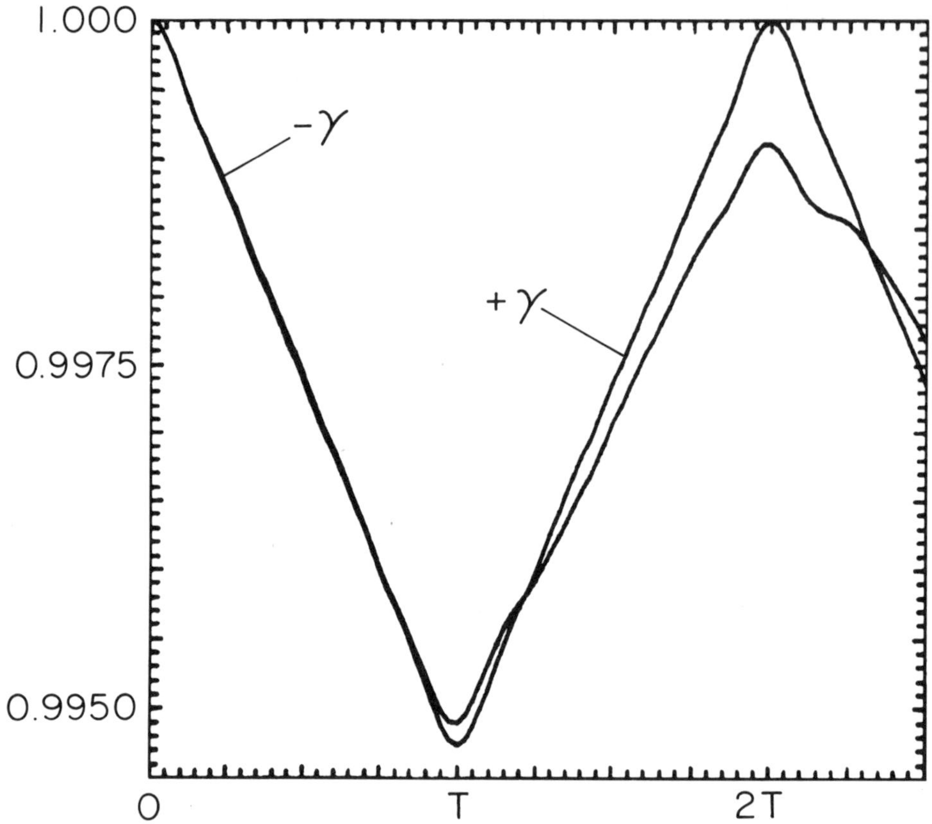

Figure 2
Probability of remaining in the ground level vs. time, under weak laser excitation, case (-). The two curves were obtained from numerical solutions of Eqs. (17), using in one case 11 equally-spaced QC levels and in the other case 11 levels that had Gaussian distributed random spacings. (The ratio of standard deviation to mean level spacing was 1/20.) The slopes and abrupt changes in slope predicted by the ICG theory are well-matched by both curves.

frequency resolution of energy levels in a time shorter than the inverse of the frequency gap between levels. What is unexpected is that the "continuum" behavior persists for a significant period beyond the initial instant after laser excitation.

We introduce a characteristic time T_{QC} associated with the QC via the density in frequency of the QC levels, namely $\rho(\omega)$. If the QC has a smooth (in a coarse-grained sense) density of levels, then $\rho(\omega) \cong$ constant, and we define

$$T_{QC} = [\rho]^{-1} \ . \tag{22}$$

We have shown that continuum behavior persists until $t = T_{QC}$. More to the point, we have also shown that the molecule's pre-dissociative behavior changes abruptly, rather than smoothly, at $t = T_{QC}$. However, immediately after this characteristic time, the molecule's dynamical response can again be accurately estimated via coarse-graining, i.e., via continuum-type calculations. In effect, there is a momentary kink in the molecular response.

In this paper we explore two issues left open in our earlier reports:

(1) How abrupt is the kink? Does it have a characteristic duration of its own, albeit much shorter than T_{QC}?

(2) Is is possible that our method of so-called "interrupted coarse-graining" is itself the cause of the kink in the analytic calculations? If this were so, it would re-open the question of the origin of the king, which is observed reproducibly in exact numerical calculations.[6] We will deal with issue #2 first, and show that our method does not introduce kinks spuriously.

INVARIANCE OF THE DYNAMICS TO ARBITRARY INTERRUPTION

If our interruption procedure[1,2] is well-founded, it should be valid for arbitrary interruption times τ, $\tau < T_{QC}$ as well as $\tau = T_{QC}$. We show in this section that this is indeed the case.

The coarse-grained solution of Eqs. (17) for $0 \leq t < T_{QC}$, is easily shown to give familiar Weisskopf-Wigner decay of the initial state:

$$a_o(t) = e^{-\kappa t} . \tag{23a}$$

The excited amplitudes have correspondingly simple solutions:

$$a_m(t) = -\frac{i\beta_m}{(i\Delta_m-\kappa)} \quad e^{-\kappa t} - e^{-i\Delta_m t} . \tag{23b}$$

Here we have defined $\kappa = K(0) = \gamma/2 + i\bar{\Delta}$, where γ and $\bar{\Delta}$ have the usual coarse-grained forms:

$$\gamma = 2\pi|\beta|^2\rho \tag{24a}$$

$$\bar{\Delta} = -P \int d\Delta\rho(\Delta)|\beta|^2/\Delta . \tag{24b}$$

We now interrupt the dynamical evolution at the arbitrary time $t = \tau$, $\tau < T_{QC}$, so solutions (23) become

$$a_o(\tau) = e^{-\kappa\tau} \tag{25a}$$

$$a_m(\tau) = -\frac{i\beta_m}{(i\Delta_m-\kappa)} \quad e^{-\kappa\tau} - e^{-i\Delta_m\tau} . \tag{25b}$$

Our method of interrupted coarse-graining then uses Eqs. (25) as initial conditions in Eqs. (17). The result for the ground state amplitude at a later time is:

$$a_o(t) = a_o(\tau)e^{-\kappa t} - i \sum_{m=1}^{N} \beta_m^* a_m(\tau)\left\{\frac{e^{-\kappa t}}{(i\Delta_m-\kappa)} + \frac{e^{-i\Delta_m t}}{(\kappa-i\Delta_m)}\right\} \tag{26}$$

for $\tau \leq t \leq T_{QC}$. Equation (26) is simplified by coarse-graining, i.e., by the replacement $\sum_m \to d\Delta\rho(\Delta)$, and we obtain

$$a_o(t) = a_o(\tau)e^{-\kappa t} - \int \frac{d\Delta\rho(\Delta)|\beta(\Delta)|^2}{(\Delta+i\kappa)^2} [e^{-\kappa\tau} - e^{-i\Delta\tau}] [e^{-\kappa t} - e^{-i\Delta t}] . \quad (27)$$

The last term in (27) is easily seen to give zero contribution for $t < T_{QC}$. Thus, (27) yields

$$a_o(t) = a_o(\tau) e^{-\kappa t} = e^{-\kappa(t+\tau)} , \quad (28)$$

$0 \leq t + \tau < T$. Equation (28) is indeed identical in form to (23a). Thus the dynamical evolution is unaffected by our analytical interruption technique when the interruption time τ is chosen arbitrarily and not equal to the characteristic time T_{QC}. The interruption procedure "works" because the system always retains phase memory even though it behaves in a coarse-grained "irreversible" manner throughout the "interior" of each characteristic interval, 0-T_{QC}, T_{QC}-$2T_{QC}$, etc.

EFFECT OF THE TOTAL BANDWIDTH ON THE (-) CASE

The previous analysis was conducted under the assumption that the bandwidth of the QC was very large. Here we examine the effect of a finite bandwidth on the so-called (-) case, i.e., the case in which the laser tuning is such that $\exp(i\Delta_m T_{QC}) \cong -1$ for all m. Since the dynamical behavior of the system is determined by the kernel $K(\tau)$, we analyze its bandwidth dependence. We consider (21) again, therefore, with uniform coupling.

$$K(\tau) = |\beta|^2 \sum_{m=1}^{N} e^{-i\Delta_m\tau} . \quad (29)$$

For the (-) case, the laser frequency ω_L is tuned midway between two levels at band center. If there are just two levels of separation ρ^{-1}, (29) yields

$$K(\tau) = 2|\beta|^2 \cos \tfrac{1}{2} \tau/T_{QC} . \quad (30)$$

The width of this function is large and the system certainly does not behave according to Fermi's Golden rule for any appreciable part of the range $0 \leq t \leq T_{QC}$.

For the case of N levels, N even, we have

$$K(\tau) = 2|\beta|^2 \sum_{p=1}^{N/2} \cos (2p-1) \frac{\tau}{2T_{QC}} \quad (31)$$

which sums to

$$K(\tau) = |\beta|^2 \frac{\sin (N \frac{\tau}{2T_{QC}}}{\sin (\frac{\tau}{2T_{QC}})} . \quad (32)$$

According to (32)

$$K_{max} = |\infty|^2 \lim_{\tau\to 0} \frac{\sin (N \frac{\tau}{2T_{QC}})}{\sin (\frac{\tau}{2T_{QC}})} = N|\beta|^2 , \quad (33)$$

$$K_{min} = 0 \text{ when } \tau = \frac{2\pi T_{QC}}{N} . \tag{34}$$

So, from (34) the first minimum of $K(\tau)$ occurs for a time inversely proportional to N (note that $T_{QC} = 2\pi\,\rho$ is held constant, i.e., fixed level separation). Thus, as the number of levels N increases, increasing the total bandwidth $\Gamma = \rho^{-1}(N-1)$, the width of the main contribution from $\kappa(\tau)$ narrows and we expect the turn-around time in the neighborhood of the recurrence time to diminish in proportion. This is shown to be the case in Fig. 3 which is the results of numerical calculations for this case for a sequence of different total bandwidths.

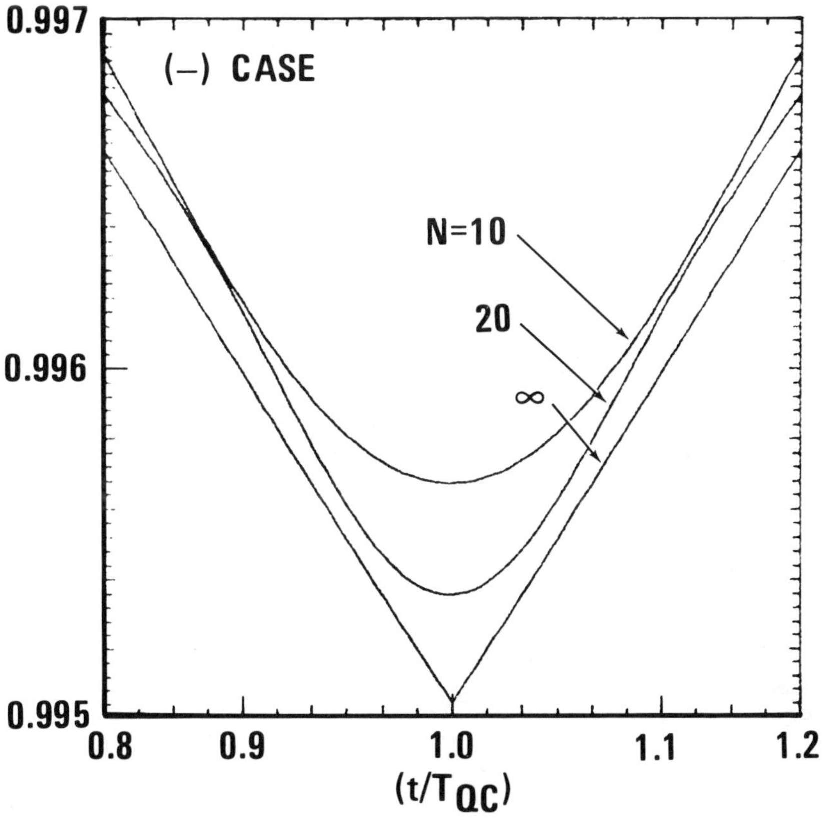

Figure 3
Probability of remaining in the ground level vs. time, under weak laser excitation, case (-). The curves were obtained from numerical solutions of Eqs. (17), using equally-spaced levels as indicated. The set of parameters for each of the curves are identical, including level spacing δ, except that the number of QC levels, i.e., width of the QC band, is varied as indicated. Approximate inverse proportionality of the turn-around time with width of the QC band as predicted in Section 5 is clearly indicated.

CONCLUSIONS

We have shown in Section 4 that the dynamics of the system portrayed in Fig. 2 are invariant to arbitrary interruption in our interrupted coarse-graining procedure.[1,2] We have also demonstrated that the turn-around time in the neighborhood of a stimulated recurrence is essentially inversely proportional to the QC bandwidth Γ as shown in Fig. 3.

ACKNOWLEDGMENTS

We have enjoyed many conversations with Dr. J. J. Yeh, and thank him for his contributions to our early understanding of interrupted coarse-graining. Discussions with Dr. B. W. Shore of Lawrence Livermore National Laboratory, and information from Drs. J. R. Ackerhalt, H. W. Galbraith, and P. W. Milonni about their related work at Los Alamos National Laboratory, have been valuable. We thank E. Kyrola for essential assistance with numerical computations, and for the production of the graphs. This work was partially supported by the Allied Corporation and Battelle-Columbus Laboratories.

REFERENCES

[1] J. H. Eberly, J. J. Yeh, and C. M. Bowden, Chem. Phys. Lett. 86, 76 (1982).

[2] J. J. Yeh, C. M. Bowden and J. H. Eberly, J. Chem. Phys. 76, 5936 (1982).

[3] The term "quasicontinuum" appears to have been first used in the context of laser-induced dissociation of polyatomic molecules in: N. R. Isenor, V. Merchant, R. S. Hallsworth, and M. C. Richardson, Can. J. Phys. 51, 1281 (1973). A recent paper with a variety of references is: H. Galbraith, J. R. Ackerhalt, and P. W. Milonni, J. Chem. Phys. (in press, 1982). For a summary of theoretical developments, see the review: C. D. Cantrell, V. S. Letokhov, and A. A. Makarov, in Coherent Nonlinear Optics, Recent Advances, edited by M. S. Feld and V. S. Letokhov (Springer, Heidelberg, 1980), Chap. 5. For early work, see, for example, N. Bloembergen, C. D. Cantrell, and D. M. Larsen, in Tunable Lasers and Applications, edited by A. Mooradian, T. Jaeger, and P. Stokseth (Springer, Heidelberg, 1976); S. Mukamel and J. Jortner, J. Chem. Phys. 65, 5204 (1976); J. Stone, M. F. Goodman, and D. A. Dows, ibid, 65, 5052; 5062 (1976).

[4] L. Allen and J. H. Eberly, Optical Resonance and Two-level Atoms, (Wiley, New York, 1975), Chap. 2.

[5] R. Gilmore, Lie Groups, Lie Algebras and Some of Their Applications, (Wiley, New York, 1974), Chap. 6, Sec. 2. R. Gilmore, C. M. Bowden and L. M. Narducci, Phys. Rev. A12, 1019 (1975); in Quantum Statistics and the Many-Body Problem, edited by S. B. Trickey, W. P. Kirk and J. W. Dufty (Plenum, New York, 1975), p. 249.

[6] Private communications from B. W. Shore and P. W. Milonni. See also, B. W. Shore and J. H. Eberly, in Lasers '82, Proceedings of International Conference on Lasers '82, New Orleans, December 1982, where the effect of the discontinuities is studied in the context of light propagation through a material with a QC band of excited levels.